T0136837

Signals and Communication Technology

Series Editors

Emre Celebi, Department of Computer Science, University of Central Arkansas, Conway, AR, USA

Jingdong Chen, Northwestern Polytechnical University, Xi'an, China

E. S. Gopi, Department of Electronics and Communication Engineering, National Institute of Technology, Tiruchirappalli, Tamil Nadu, India

Amy Neustein, Linguistic Technology Systems, Fort Lee, NJ, USA

H. Vincent Poor, Department of Electrical Engineering, Princeton University, Princeton, NJ, USA

This series is devoted to fundamentals and applications of modern methods of signal processing and cutting-edge communication technologies. The main topics are information and signal theory, acoustical signal processing, image processing and multimedia systems, mobile and wireless communications, and computer and communication networks. Volumes in the series address researchers in academia and industrial R&D departments. The series is application-oriented. The level of presentation of each individual volume, however, depends on the subject and can range from practical to scientific.

"Signals and Communication Technology" is indexed by Scopus.

More information about this series at http://www.springer.com/series/4748

Fernando Gregorio • Gustavo González
Christian Schmidt • Juan Cousseau

Signal Processing Techniques for Power Efficient Wireless Communication Systems

Practical Approaches for RF Impairments Reduction

 Springer

Fernando Gregorio
Dpto. de Ing. Eléctrica y de Computadoras
Universidad Nacional del Sur (UNS),
Instituto de Inv. en Ing. Eléctrica "Alfredo
Desages" (IIIE), UNS-CONICET
Bahía Blanca
Buenos Aires, Argentina

Gustavo González
Dpto. de Ing. Eléctrica y de Computadoras
Universidad Nacional del Sur (UNS),
Instituto de Inv. en Ing. Eléctrica "Alfredo
Desages" (IIIE), UNS-CONICET
Bahía Blanca
Buenos Aires, Argentina

Christian Schmidt
Dpto. de Ing. Eléctrica y de Computadoras
Universidad Nacional del Sur (UNS),
Instituto de Inv. en Ing. Eléctrica "Alfredo
Desages" (IIIE), UNS-CONICET
Bahía Blanca
Buenos Aires, Argentina

Juan Cousseau
Dpto. de Ing. Eléctrica y de Computadoras
Universidad Nacional del Sur (UNS),
Instituto de Inv. en Ing. Eléctrica "Alfredo
Desages" (IIIE), UNS-CONICET
Bahía Blanca
Buenos Aires, Argentina

ISSN 1860-4862   ISSN 1860-4870   (electronic)
Signals and Communication Technology
ISBN 978-3-030-32439-1   ISBN 978-3-030-32437-7   (eBook)
https://doi.org/10.1007/978-3-030-32437-7

This Springer imprint is published by the registered company Springer Nature Switzerland AG.
The registered company address is: Gewerbestrasse 11, 6330 Cham, Switzerland

To Patricia, Manuela, Delfina, and Luisa.

Fernando

To my family and friends.

Gustavo

To my friends, my family, and Marina.

Christian

To Cristian, Ignacio, and Miriam.

Juan

Preface

This book is a synthesis of the tasks and research experiences carried out in the Laboratory of Signal Processing and Communications (LaPSyC), CONICET, Universidad Nacional del Sur, Argentine, since 2003.

It includes broadly known models and techniques, used in the signal processing community, which facilitates their application and understanding.

Emphasis is placed on methodologies of low complexity and scalable to different applications. It also emphasizes measures of performance and impact of each compensation technique in the corresponding context.

To allow a more direct reading, not only for academic community but also for industry, the contents are divided into three parts: (1) basic models, (2) compensation techniques, and (3) applications in advanced technologies.

The first part describes, on the one hand, basic architectures of transceivers and their component blocks, and on the other, modulation techniques. Together they describe the behavior to be taken into account, regardless of the distortions to compensate produced by the specific technologies of implementation.

In the second part, several schemes of compensation and/or reduction of imperfections are addressed, among which we highlight: linearization of power amplifiers, compensation of the characteristic of analog-to-digital converters, and CFO compensation for OFDM modulation.

In the third part, some of the previous techniques are used in modern applications of wireless communications systems such as full-duplex transmission, massive MIMO schemes, and finally applications in internet of things.

Bahía Blanca, Argentina
Bahía Blanca, Argentina
Bahía Blanca, Argentina
Bahía Blanca, Argentina
July 2019

Fernando Gregorio
Gustavo González
Christian Schmidt
Juan Cousseau

Contents

Acronyms

3

3GPP   Third generation project partnership

5

5G   Fifth Generation

6

6LoWPAN IPv6 protocol over Low power Wireless Personal Area Networks

A

ACB	Access class barring
ACLR	Adjacent channel power leakage ratio
ACPR	Adjacent channel power ratio
ADC	Analog to digital converter
AF	Amplify-and-forward
AFE	Analog frond-end
AGC	Automatic gain control
ALOHA	Medium access techniques amenable to Wi-Fi and mobile networks
AM/AM	Amplitude to amplitude conversion
AM/PM	Amplitude to phase conversion

AWGN   Additive white Gaussian noise

B

BC   Banded compensation
BER   Bit error rate
BLUE   Best linear unbiased estimator

C

CAS   Carrier assignment scheme
CBC   Circular banded compensation
CDMA   Code division multiple access
CF   Compress-and-forward
CFO   Carrier frequency offset
CIR   Channel impulse response
CP   Cyclic prefix
CPE   Common phase error
CT   Continuous-time

D

DAC   Digital-to-analog converter
DF   Decode-and-forward
DFT   Discrete Fourier transform
DINL   Differential nonlinearity
DL   Downlink
DMRS   Demodulation reference signal
DPD   Digital predistorter
DRX   Discontinuous receiving mode
DSSS   Direct sequence spread spectrum
DTVM   Discrete-time Volterra model
DUT   Device under test

E

EC-GSM   Extended coverage GSM

EE	Energy efficiency
eMBB	Enhanced mobile broadband
ENOB	Effective number of bits
EVM	Error vector magnitude

F

FBMC	Filter bank based multicarrier
FD	Full duplex
FDD	Frequency-division duplexing
FFT	Fast Fourier transform
FHSS	Frequency hopped spread spectrum

G

GCAS	Generalized CAS
GFSK	Gaussian frequency shift keying
GSM	Global system for mobile communications

H

H2H	Human to human communications
HARQ	Hybrid automatic repeat request
HD	Half-duplex
HONT	High-order noise terms

I

IBO	Input back-off
ICAS	Interleaved CAS
ICI	Intercarrier interference
IEEE	Institute of electrical and electronics engineers
IETF	Internet engineering task force
INL	Integral nonlinearity
IP3	Third order interception point
IPN	Integrated phase noise
I/Q	In-phase and quadrature

IRR	Image rejection ratio
ISI	Intersymbol interference
ISM	Industrial scientific and medical

K

KPI	Key performance parameters

L

LNA	Low noise amplifier
LO	Local oscillator
LoRa	Long range technology for machine-type communications
LPWA	Global system for mobile communications
LSB	Least significant bit
LTE	Long term evolution
LTE-M	(LTE-MTC) Machine type communications LTE
LUT	Look-up table

M

MAI	Multiple access interference
MCL	Maximum coupling loss
MIB	Master information block
MLE	Maximum likelihood estimator
MMSE	Minimum mean square error
mMTC	Massive machine-type communications
MPL	Maximum path loss
MSE	Mean square error

N

NB-IoT	Narrowband Internet of things
NLC	Nonlinear canceler
NOMA	Non-orthogonal multiple access
NPBCH	Narrowband physical broadcast channel
NPDCCH	Narrowband physical DL control channel

NPDSCH	Narrowband physical DL shared channel
NPSS	Narrowband primary synchronization signal
NRS	Narrowband reference signal
NSSS	Narrowband secondary synchronization signal

O

OBO	Output back-off
OFDM	Orthogonal frequency-division multiplexing
OFDMA	Orthogonal frequency division multiple access
OQAM	Offset QAM
OSR	Over-sampling ratio

P

P-DPD	Partitioned digital predistorter
PA	Power amplifier
PANC	Power amplifier nonlinearities cancellation
PAPR	Peak-to-average power ratio
PB	Processing block
PD	Predistorter
PLL	Phase locked loop
PN	Phase noise
PRB	Physical resource block
PSD	Power spectral density
PSM	Power saving mode

Q

QAM	Quadrature amplitude modulation
QoS	Quality of service

R

RF	Radio frequency
RRC	Radio resource control
RX	Receiver

S

SAR	Succesive approximation register
SC-FDMA	Single carrier frequency division multiple access
SCAS	Subband CAS
SDM	Sigma-delta modulator
SE	Spectral efficiency
SFDR	Spurious free dynamic range
SH	Sample and hold
SI	Self interference
Sigfox	Global network operator oriented to machine-type communications
SINAD	Signal-to-noise and distortion ratio
SINR	Signal to interference and noise ratio
SL	Soft limiter
SNR	Signal to noise ratio
SQNR	Signal to quantization noise ratio
SSPA	Solid state power amplifier

T

TD	Total degradation
TDD	Time-division duplexing
TI	Time-interleaved
TX	Transmitter

U

UL	Uplink
URLLC	Ultra reliable and low latency communications

W

WCDMA	Wideband code division multiple access

Z

Zigbee   Open global standard designed to use low-power digital radio signals for personal area networks

Part I
Definitions and Models

This part is composed of three chapters. Chapter 1 motivates the topics of the book. Chapter 2 introduces digital block and RF front-end models. Important design metrics are defined. Also, wireless channel, power amplifier, low noise amplifier, mixers, ADC, and DAC models are discussed in detail. Chapter 3 is focused on aspects of energy consumption in transceiver design. To that purpose, energy efficiency and spectral efficiency are defined. Also, modeling of power driven parts of the digital block, the front-end, and the ADC are considered. From the perspective of power balance, different (short-range and long-range) links and power scaling are contemplated.

Chapter 1
Introduction

Abstract This chapter aims to motivate the topics of the book. A general discussion about fifth generation (5G) wireless systems and its requirements is presented first. That is useful to put in evidence that the main objective of the book is related to modern wireless transceiver design (rather than on system design). From that perspective, the basic blocks (front-end, baseband processing, and analog-to-digital conversion) of a generic transceiver and its design parameters are discussed. The main contributions of the book are addressed to study the impact of RF impairments in 5G transceiver designs and also to introduce successful estimation and compensation techniques.

1.1 Motivation: 5G Wireless Systems and Its Requirements

The wireless mobile telephony industry has made considerable progress in recent years given that the LTE standard has brought together a large part of the mobile phone industry. In certain regions of the world, LTE implementations continue to expand to cover countries such as Korea, Japan, China, and the USA [1]. This has led the mobile industry to focus on 5th generation (5G) mobile technology. The new technologies take into account the growing demand for higher performance and greater data capacity, particularly for video, to provide better broadband services through enhanced mobile broadband communications. In addition, the goal of 5G is to address new markets and applications, such as mass machine-type communications and ultra-low-latency communications, among others.

Given the continued focus on LTE deployment and the parallel industry effort to define 5G, 3rd Generation Partnership Project (3GPP) has been actively working in both tracks. With additional detail for the applications in mind, 5G requirements regard the following cases [2]:

- Enhanced mobile broadband (eMBB): The scenario addresses different service areas (e.g., indoor/outdoor, urban and rural areas, office and home, local and wide areas connectivity), and special deployments (e.g., massive gatherings, broadcast, residential, and high-speed vehicles). For instance, for the downlink, experienced

© Springer Nature Switzerland AG 2020
F. Gregorio et al., *Signal Processing Techniques for Power Efficient Wireless Communication Systems*, Signals and Communication Technology,
https://doi.org/10.1007/978-3-030-32437-7_1

data rate of up to 50 Mbps are expected outdoor and 1 Gbps indoor (5GLAN), and half of these values for the uplink.

- Ultra reliable and low latency communications (URLLC): Several scenarios require the support of very low latency and very high communications service availability. These are driven by the new services such as industrial automation. For instance, in the context of remote control for process automation, a reliability of 99.9999% is expected, with a user experienced data rate up to 100 Mbps and an end-to-end latency of 50 ms.
- Massive machine-type communications (mMTC). Several scenarios require the 5G system to support very high traffic densities of devices. The mMTC requirements include the operational aspects that apply to the wide range of internet of things (IoT) devices and services anticipated in the 5G plans.

Figure 1.1 depicts the expected requirements (weak to strong requirements from the center to the edges) for the mentioned applications (eMBB, URLLC, and mMTC) in terms of important design dimensions such as: coverage, number of supported connections, latency, throughput, mobility, device complexity, and device battery life.

This diversity of requirements leads to different key performance parameters (KPI) that are used to define specific 5G system and transceiver design solutions. Considering that the main emphasis in this book is on transceiver design, we describe in the following some basic concepts and notation about modulation and transceiver parts. They will be useful to introduce the objectives and outline of the book in the final sections of this chapter.

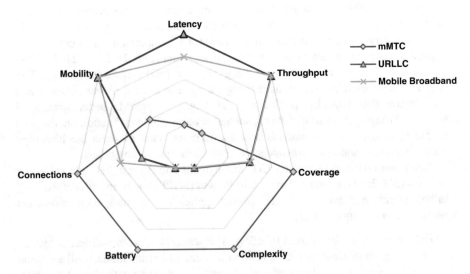

Fig. 1.1 5G requirements for different application cases

1.2   Basic Components of the Communication System

Taking into account the diversity of requirements envisioned for 5G wireless mobile system, a wide range of technologies are considered for spectrum access (enhanced carrier aggregation schemes, integrated license-exempt spectrum, use of frequencies beyond 6 GHz, full-duplex), radio link (new waveforms, NOMA, different radio frame numerology, massive MIMO and multiantenna schemes, NB-IoT), and radio access capacity (densification, dual connectivity, enhanced radio access technology, D2D, etc.)

Throughout the book, we are going to emphasize aspects related to the radio link design, most of them specified in the 3GPP rel. 15 [2]. These aspects are, logically, closely related to each component of the transceiver design problem.

A key design aspect of a typical transceiver is the modulation format to be adopted. Owing to their high data throughput, spectral efficiency, and versatility, the most promising techniques for the new generation of wireless systems are based on multicarrier modulation schemes. Two candidates for the uplink are orthogonal frequency division multiple access (OFDMA) and filter bank based multicarrier (FBMC) [3, 4].

OFDMA is an extension of OFDM, considering the implementation of multiuser communication systems. A subset of subcarriers are assigned to each user according to a carrier allocation scheme (CAS). The usual allocation schemes are subband, interleaved, and generalized [5]. In subband CAS, each user takes a contiguous set of subcarriers. In interleaved CAS, the carriers of each user are uniformly distributed over the entire signal bandwidth to exploit the frequency diversity. Nevertheless, the more advantageous scheme is generalized CAS since it allows users to be allocated to the best currently available subcarriers [5].

FBMC transmission technique, on the other hand, can be also thought as an extension of the OFDM concept, replacing the rectangular time window by a general impulse response. After parallelization, each symbol is extended using that impulse response producing a time domain overlapping [4]. The transmitter/receiver filter pair is derived from the Nyquist criterion in order to avoid intersymbol interference. The time extension produces highly selective filters that lead to reduced intercarrier interference (ICI) and also to reduced multiple access interference (MAI).

A diagram of a generic wireless transceiver is illustrated in Fig. 1.2. Three basic blocks can be used to describe a generic transmitter (TX)/receiver (RX): the analog block, the digital block, and their interface. The analog block refers to the RF TX/RX front-end, usually associated with antenna array (TX and/or RX), TX power amplifier, RX low noise amplifier and possibly mixers and/or conditioning filters. The digital block is related to all baseband signal processing of the adopted waveform signaling. That includes (from a very simplistic perspective): coding, interleaving, modulation at the transmitter, in addition to diversity management, synchronization, and decoding at the receiver. Between these blocks it is necessary to perform the corresponding map among signals and symbols from one domain

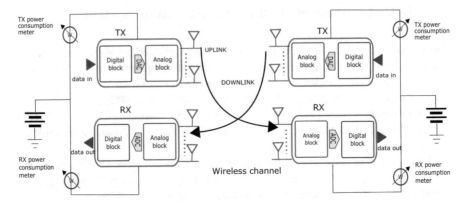

Fig. 1.2   Generic wireless transceiver (downlink/uplink)

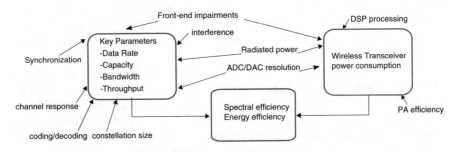

Fig. 1.3   Illustration of figures of merit evaluation and their dependence with TX/RX parameters

to the other using a digital-to-analog converter (DAC) and an analog-to-digital converter (ADC).

Important parts of the design process of a wireless transceiver (composed of these blocks) are summarized in Fig. 1.3. We consider the uplink (UL)/downlink (DL) specifications on one side (data rate, bandwidth, throughput, capacity), and the spectral efficiency and power consumption as key objectives. There are several interrelationships between the three blocks, but it is possible to put in evidence some specific aspects particular to each one. When considering the UL/DL design, some specific compromise about coding/decoding, constellation size, channel model, and synchronization aspects must be sought. Spectral efficiency is directly related to the selection of these parameters. From the perspective of power consumption, TX power amplifier efficiency (and also baseband processing, depending of the specific application) need to be considered. Furthermore, DAC/ADC resolution has a considerable impact in the transceiver design. Finally, not trivial impairments (as TX PA nonlinearities, ADC performance, interference, etc.) must be contemplated in a generic design.

1.3 Implementation Issues

When considering the three basic blocks composing a generic transceiver, non-trivial impairments and non-ideal models must be taken into account to obtain a reasonable and practical design. An illustration of possible impairments to be considered is given in Fig. 1.4.

Up- and down-converters (mixers) must be compensated for in-phase and quadrature branches (I/Q) imbalance, since that impairment leads to constellation degradation, effective constellation size reduction, and data rate reduction. Carrier frequency offset (related not only to low quality RX oscillator but also to channel Doppler) must be considered too due to the intercarrier interference and signal-to-noise ratio reduction it produces. Furthermore, phase noise associated with local oscillators produces intercarrier interference, bit error rate degradation, and additional signal to noise ratio (SNR) reduction. Power amplifier (nonlinear) distortion produces low power efficiency and generates out-of-band interference. Among the impairments to be considered, ADC/DAC non-ideal behavior is also very important since it produces high quantization errors, increases power consumption, and introduces nonlinear distortion.

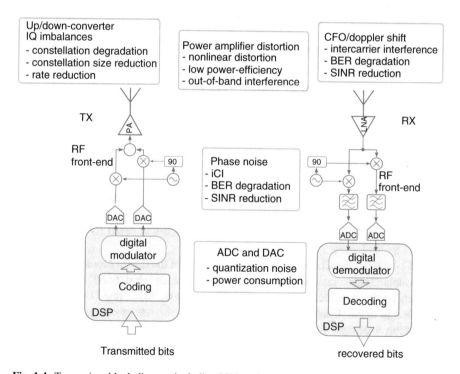

Fig. 1.4 Transceiver block diagram including RF impairments and their effects

All these impairments will be considered, and their impact discussed, in different design methodologies throughout the book, as explained in the following sections.

1.4   Main Contributions of the Book

The main contributions of the book are aimed to put models and methodologies useful for last generation wireless transceiver design in the hands of a broad audience. The material collected and organized corresponds to several years of experiences and studies of the research group working at the Signal Processing and Communications Laboratory (Laboratorio de Procesamiento de Señales y Comunicaciones, LaPSyC), Department of Electrical and Computer Engineering, Universidad Nacional del Sur, Bahía Blanca, Argentina. Some parts of this work were closely related to the PhD thesis dissertation topics of the authors and other parts were motivated when considering different applications.

Specifically, the subjects where novel contributions are made can be summarized in the following:

- Modeling RF impairments and their effects over energy and spectral efficiency.
- Digital based RF compensation techniques: implementation cost and energy saving effects.
- RF impairments on novel technologies, i.e., their impart on massive MIMO, full-duplex, and IoT nodes.

The way these contributions are developed throughout the book is introduced in the following section.

1.5   Outline of the Book

Coherently with the main contribution topics, the book is divided into three parts: Part I mainly addressed to definitions, notation, and models; Part II to introduce digital compensation techniques; and Part III to discuss in detail RF imperfections in novel technologies.

Part I continues in Chap. 2 with the introduction of the digital block and RF front-end models. There, basic but important design metrics are defined. Also, wireless channel, power amplifier, low noise amplifier, mixers, ADC, and DAC models are discussed in detail. Chapter 3 is devoted to discuss aspects of energy consumption in transceiver design. To that purpose, energy efficiency and spectral efficiency are defined. Also, modeling of power driven parts of the digital block, the front-end, and the ADC are considered. From the perspective of power balance, different (short range and long range) links and power scaling are contemplated.

Part II of the book begins with Chap. 4, where power amplifier linearization techniques are presented. In- and out-of-band distortion are parameterized with figures of merit to measure degrees of nonlinearity. Basic transmitter-side predistortion techniques, aiming linearization, and receiver side compensation techniques are also introduced. Finally, a case of study (class AB and envelope-tracking PA) is introduced in this chapter. Chapter 5 studies different classes of ADC when aiming to discuss the trade-off between conversion speed and consumption: flash, successive approximations, sigma-delta, and combined structures. Considering the possible modeled impairments (integral nonlinearity and dynamic nonlinear models), different compensation techniques are introduced: model inversion, mismatch error compensation, etc. This part concludes with Chap. 6, where frequency offset and phase noise effects are described and discussed. Specific cases where these impairments are critical are discussed in detail. Also, estimation and compensation of carrier frequency offset in the DL and UL is considered.

Part III of the book starts with Chap. 7, where full-duplex (FD) transceiver design aspects are introduced. Feasible FD design requires different techniques to be used together. The main emphasis here is put on baseband signal processing for self-interference reduction. An important aspect to be considered is the modelling of the effects that the non-ideal behavior of the basic transceiver blocks produce in the system performance. Chapter 8 introduces massive MIMO systems. DL and UL aspects are considered in addition to channel non-reciprocity, antenna coupling, channel estimation, and pilot contamination. Minimum RF requirements (mostly ADC resolution) are key to the front-end design in this case and are discussed in detail, including a numerical evaluation. Also, power consumption design aspects are considered. This part of the book also includes Chap. 9, where machine-type communications are studied. More specifically, internet of things applications and challenges are discussed. Main design concepts and parameters for IoT design are introduced. Emphasis is made on licensed IoT standards, mainly narrowband IoT (NB-IoT) [2]. In addition, a study of RF impairments effects from LTE on NB-IoT is included. Finally, Part III concludes with Chap. 10 where some 5G implementation challenges are discussed, among them: combination of massive MIMO and FD techniques, millimeter wave wireless design, massive MIMO challenges, and non-orthogonal multiple access schemes.

References

1. Next Generation Mobile Networks Ltd, *A Deliverable by the NGMN Alliance, NGMN 5G White Paper* (2015)
2. 3rd Generation Partnership Project (3GPP), *Technical Specification Group Services and System Aspects; Release 15 Description; Summary of rel-15 Work Items, 3GPP TR 21.915* (2019)
3. H. Holma, A. Toskala, *LTE for UMTS–OFDMA and SC-FDMA Based Radio Access*, 1st edn. (Wiley, New York, 2009)
4. PHYDYAS deliverable D2.1: data-aided synchronization and initialization (single antenna), in *Physical Layer for Dynamic Access and Cognitive Radio* (2009)
5. M.-O. Pun, M. Morelli, C.C.J. Kuo, *Multi-Carrier Techniques for Broadband Wireless Communications: A Signal Processing Perspectives* (Imperial College Press, London, 2007)

Chapter 2
Digital Block and RF Front-End Models

Abstract In order to reduce the implementation cost, most of the signal processing in the transceiver is carried out in the digital domain. However, the RF front-end components usually represent a significant part of the cost and power consumption, and determine the overall performance of the radio system. In this chapter, we present a review of the most significant front-end impairments that affect the performance of modern wireless communication systems. At the transmitter side, we consider power amplifier (PA) nonlinear distortion and the phase and amplitude imbalances of the mixer. At the receiver, we include the phase noise of the local oscillator, and the analog-to-digital converter (ADC) quantization noise and nonlinear distortion. These models will be used in the following chapters to introduce compensation techniques that improve system performance.

2.1 Introduction

Novel communication systems, particularly massive market implementations, impose several requirements as low cost, low power consumption, reduced size, and reconfigurability. In addition, the requirements of large spectral efficiency and robustness against time dispersive channels motivate the use of multicarrier modulation techniques. In this context, orthogonal frequency division multiplexing (OFDM) is adopted in the majority of wireless standards, and is also one of the main candidates for 5G communication systems [1]. That modulation provides large spectral efficiency, excellent performance in time-dispersive channels, and support multiuser communication. However, despite of these advantages, OFDM is easily affected by the radio frequency impairments associated with the analog front-end [2].

Several RF imperfections degrade the system performance, e.g., inaccurate local oscillators, mismatches in the phase (I) and quadrature (Q) branches of the baseband conversion, and nonlinear power amplifiers. Furthermore, low noise amplifiers (LNA) and variable gain amplifiers with nonlinear response, analog-to-

© Springer Nature Switzerland AG 2020
F. Gregorio et al., *Signal Processing Techniques for Power Efficient Wireless Communication Systems*, Signals and Communication Technology, https://doi.org/10.1007/978-3-030-32437-7_2

digital (ADC) and digital-to-analog (DAC) conversion with low resolution are also typical degradation sources of communication transceivers [3, 4].

In order to quantify the degradation introduced by each imperfection on the system performance, specific figures of merit are introduced. Error vector magnitude (EVM) and signal-to-interference-plus-noise ratio (SINR) provide information regarding the distortion introduced at the transmitter side and the receiver side, respectively.

A block diagram of a typical wireless transceiver is illustrated in Fig. 2.1 where the whole system is depicted. Two essential blocks are included in the transceiver: the baseband digital processing block and the (transmitter/receiver) RF front-end, interfaced by ADCs and DACs.

This chapter begins with a description and characterization of the digital processing block. A mathematical model of the baseband multicarrier modulation technique, particularly OFDM, and typical parameters are presented in Sect. 2.4. After the description of the digital block, modeling aspects of the complete RF front-end are introduced, including mathematical models of the most significant components of the RF front-end, such as the ADC, DAC, mixer, LNA, and the power amplifier.

Fig. 2.1 Block diagram of communication transceiver

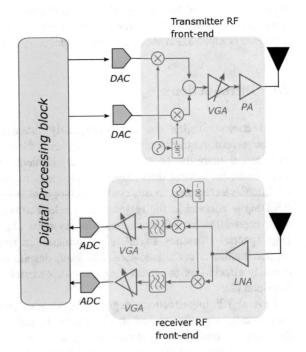

2.2 Metrics for Wireless Communication Systems

The design of a wireless communication system must contemplate a good balance in the selection of its components. A low noise amplifier with good performance in terms of noise figure and gain, mixers with large conversion gain and high isolation between RF and baseband signals, ADC with an adequate dynamic range and low quantization noise, energy-efficient and linear power amplifiers, and stable local oscillators are typical requirements of RF front-end design.

On the other hand, it is necessary to satisfy the tight constraints on the power consumption, implementation cost, and size of the individual radios.

As a consequence, technologies to reduce the size and cost of massive-market products are employed [5]. The use of low-cost components creates several imperfections that limit the system performance. Usually, a certain level of imperfection in the front-end is allowed that can be compensated at the digital domain if necessary. The trade-off between implementation complexity of digital compensation techniques and the cost of analog front-end has to be carefully evaluated [6]. In this section, several metrics to evaluate the quality of a transmitter and receiver are described. At the end of the section, a short review of channel characterization, including deterministic and statistical models, is presented.

2.2.1 Link Budget

The operation of a communication system requires an adequate level of signal-to-noise ratio (SNR) at the reception antenna to support the chosen modulation scheme. The SNR is defined by

$$SNR = \frac{P_{RX}}{P_n},\qquad(2.1)$$

where P_{RX} is the power at the reception antenna and P_n is the received noise power.

Considering the communication system as depicted in Fig. 2.2, the link budget defines the requirements in terms of transmitted and received power, and equipment characteristics [7]. The power at the reception antenna expressed in dB is given by

$$P_{RX} = P_{TX} - PL + G_{TX} + G_{RX},\qquad(2.2)$$

where P_{TX} is the transmitted power, PL is the channel path loss, and G_{TX} and G_{RX} are the transmitter and receiver antenna gains, respectively.

The noise at the receiver can be written as:

$$P_n = P_{Tn} + 10\log(B) + 10\log(NF) + 10\log(G_{RX}),\qquad(2.3)$$

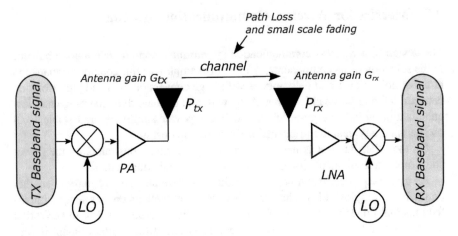

Fig. 2.2 Link budget calculation

where P_{Tn} is the thermal noise (-174 dBm for T $=$ 17 C), B is the operation bandwidth, and NF is the noise figure that quantifies the noise created at the receiver.

From the previous equations, it can be observed that the general objective of maximizing the SNR can be achieved by increasing the transmitted power and/or minimizing the noise added by the receiver. The transmitted power defines the power amplifier (PA) characteristics. The linearity and high power efficiency are the most sought features in a PA. On the other hand, the noise added at the receiver depends on the characteristics of the LNA. These specifications are discussed in the following.

2.2.2 Nonlinearities

Electronic devices are inherently nonlinear. Particularly, power amplifiers operating with large input signals present a nonlinear response which can affect the system performance by reducing the ratio between the effective desired signal and distortion [8].

The nonlinear behavior is characterized by several parameters as gain compression, inter-modulation distortion, and harmonics. The typical behavior of a memoryless time-invariant nonlinear PA, that can be modeled by a third-order polynomial, has the following expression:

$$x_{NL}(t) = a_1 x(t) + a_2 x^2(t) + a_3 x^3(t), \tag{2.4}$$

where $x(t)$ is the input signal, and a_1, a_2, and a_3 are real-valued coefficients. One-tone and two-tone input signals are employed to characterize the PA response. The output of the PA for a single-tone signal, expressed as $x(t) = C\sin(\omega_1 t)$, can be written as:

$$x_{NL}(t) = \left(a_1 C + \frac{3a_3}{4}C^3\right)\sin(\omega_1 t) + \frac{a_2}{2}C^2\sin(2\omega_1 t) + \frac{a_3}{4}C^3\sin(3\omega_1 t).$$

$$(2.5)$$

From this equation it is possible to conclude that a nonlinear device generates additional spectral components and also compress the gain at the fundamental frequency. The gain compression is characterized by the **1 dB compression point**, that is the point where the gain of a nonlinear amplifier deviates 1 dB from the gain of an ideal amplifier.

To evaluate the intermodulation products generated by the mixing of two signals, the nonlinear device is excited by a two-tone signal $x(t) = C_1\sin(\omega_1 t) + C_2\sin(\omega_2 t)$. Discarding the harmonics and DC terms, the intermodulation products are

$$x_{imp}(t) = a_2 C_1 C_2 \sin((\omega_1 + \omega_2)t) + a_2 C_1 C_2 \sin((\omega_1 - \omega_2)t)$$

$$+\frac{3}{4}a_3 C_1^2 C_2 \sin((2\omega_1 + \omega_2)t) + \frac{3}{4}a_3 C_1^2 C_2 \sin((2\omega_1 - \omega_2)t)$$

$$+\frac{3}{4}a_3 C_2^2 C_1 \sin((2\omega_2 + \omega_1)t) + \frac{3}{4}a_3 C_2^2 C_1 \sin((2\omega_2 - \omega_1)t). \quad (2.6)$$

The third-order intermodulation products at frequencies $2\omega_1 - \omega_2$ and $2\omega_2 - \omega_1$ are critical because these components appear close to ω_1 and ω_2 creating *in-band distortion*. The *interception point (IIP)* is used to quantify the intermodulation effect. Particularly, the IP3 characterizes third-order nonlinearities and defines the point where the values of the extrapolated linear gain and the distortion products are equal.

A cascade of PA devices is usually employed in a typical transceiver to increase the gain and the output power (transmitter). The third-order interception point of a cascade of N devices is given by

$$\frac{1}{IP3_o} = \left[\left[\frac{1}{IP3_N}\right]^3 + \left[\frac{1}{G_N IP3_{N-1}}\right]^3 + \cdots + \left[\frac{1}{G_N G_{N-1}\ldots G_2 IP3_1}\right]^3\right]^{1/3}.$$

$$(2.7)$$

From this equation it is easy to conclude that the last element of the cascade dominates the linearity of the system.

2.2.3 Noise Figure

The RF front-end of a receiver composed by the cascade of a LNA and a mixer is illustrated in Fig. 2.2. A good performance in several metrics as noise figure (NF), third-order intercept point (IIP3), and 1-dB compression point is required. It is worth to point out that, in general, the improvement of those metrics requires a higher power consumption.

The NF indicates the amount of additional noise generated by a device in relation to the SNR at its input. In other words, NF denotes the degradation in SNR introduced by inserting a component, and is defined as

$$NF = \frac{SNR_i}{SNR_o},\tag{2.8}$$

where SNR_i and SNR_o are the SNRs at the input and output of the device, respectively.

The noise figure of the cascade of the circuits that form the receiver front-end is an important parameter since it determines the SNR of the overall system. It can be expressed as:

$$NF = NF_1 + \frac{NF_2 - 1}{G1} + \frac{NF_3 - 1}{G_1 G_2} + \cdots + \frac{NF_n - 1}{G_1 G_2 \ldots G_{N-1}},\tag{2.9}$$

where G_n and NF_n denote the gain and noise figure of stage n (linear scale). This equation allows to conclude that when the first device has enough gain, it dominates the noise figure. For this reason, in receiver front-ends, a high gain low noise amplifier is the first component in the cascade.

In addition to the noise figure, the receiver sensitivity and dynamic range are also critical metrics to be considered. They are defined as following:

- Sensitivity: it is the minimum signal level at the antenna needed to produce a predefined signal-to-noise ratio SNR_d at the input of the detector.

  The sensitivity, expressed in dBm, can be written as:

$$P_{sen} = -174 + 10\log(B) + NF + SNR_d,\tag{2.10}$$

  where $-174 + 10\log(B) + NF$ denotes the noise floor of the receiver. The SNR_d is the minimum SNR level for what the receiver achieves the specifications in the standard.

- Dynamic range: it is the ratio of the maximum input level that the receiver can tolerate without performance degradation, to the minimum input signal level (sensitivity). The dynamic range is given by

$$DR = 10\log\left(\frac{P_{sat}}{P_{sen}}\right),\tag{2.11}$$

where P_{sat} denotes the largest RF power allowed at the receiver input. The allowed distortion depends strongly of the modulation type and the receiver architecture.

In modern communication systems that employ multicarrier modulation and high constellation size, a highly linear response is required to maximize the input signal operation and dynamic ranges.

2.2.4 Error Vector Magnitude

The degradation that each RF impairment produces on the system performance can be quantified by the error vector magnitude (EVM). It provides adequate information regarding the in-band distortion generated by all the transceiver blocks, and is widely adopted in communication standards [9], where the maximum allowable value as a function of constellation size and coding data rate is specified. Naturally, the larger the constellation size, the smaller the tolerated EVM, because the symbol decision regions are also smaller for a fixed average transmitter power. Alternatively, the knowledge of the resulting EVM for a particular RF setup determines the maximum constellation size (data rate) at which the wireless system is able to operate.

The EVM figure of merit for an OFDM system is given by

$$EVM[k] = \sqrt{\frac{E[|e[k]|^2]}{E[|X[k]|^2]}}, \qquad (2.12)$$

where $EVM[k]$ is the EVM at subcarrier k, $E[\cdot]$ denotes the expectation operator, and $e[k]$ is the error defined as the difference between the original and the received symbol after down-conversion and equalization. Since in general the interest is in the degradation of the transmitted signal, an ideal receiver is assumed, i.e., without considering any impairment, communication channel, or noise.

2.3 Wireless Channel Models

A signal transmitted through a wireless channel is received at destination along several paths which are affected by scattering, reflections, and shadowing. The variations in the power of the received signal in a wireless link are determined by several effects. These effects can be classified as a function of the spatial dimension as small-scale fading and large-scale fading.

Macroscopic or large-scale fading is related with the free-space path loss. It depends on the distance between the transmitter and receiver antennas, and can be described with a deterministic model. Considering ideal free-space propagation, the path loss is given by

$$PL = \left(\frac{\lambda}{4\pi d}\right)^2,$$ (2.13)

where d is the distance between the transmitter and receiver antennas, and λ is the signal wavelength.

On the other hand, the constructive and destructive sum of multipath components creates small-scale fading that occurs at the carrier wavelength scale. It creates fluctuations around the large-scale fading, and is characterized by a statistical model. The discrete received baseband signal $y(m)$, obtained after passing the original signal $x(m)$ through a discrete channel modeled of N_h-taps, can be represented by

$$y(m) = \sum_{l=0}^{N_h-1} h_l(m)x(m-l) + w(m),$$ (2.14)

where $w(m)$ is the channel noise and $h_l(m)$ is the time-variant discrete channel tap.

Assuming that each channel tap is the sum of independent scattered components, $h_l(m)$ can be modeled as independent zero-mean Gaussian random variables $CN(0, \sigma_l^2)$ [10]. Therefore, the magnitude of channel coefficients follows a Rayleigh density function, expressed by

$$f(y) = \frac{2y}{\sigma_l^2} \exp\left(\frac{-y^2}{2\sigma_l^2}\right).$$ (2.15)

The Rayleigh fading model is a good approximation for scenarios with large number of scatterers without any dominant direction of arrival. If the channel has a line-of-sight (LOS) component, it can be modeled as the combination of the LOS and complex Gaussian terms. The magnitude of this channel coefficient follows a Rician density distribution, and can be expressed as:

$$h_l(m) = \sqrt{\frac{K_r}{K_r+1}}\sigma_l \exp(j\theta) + \sqrt{\frac{1}{K_r+1}}CN\left(0, \sigma_l^2\right),$$ (2.16)

where the first term denotes the LOS component arriving with uniform phase θ, and the second term expresses the combination of reflected and scattered paths. The parameter K_r is called the Rice factor and denotes the ratio between the energy of the direct component to the energy in the scattered paths.

The channel including path loss (large-scale), multipath fading (small-scale), and additive noise is illustrated in Fig. 2.3.

Fig. 2.3 Channel model including small- and large-scale fading and additive noise

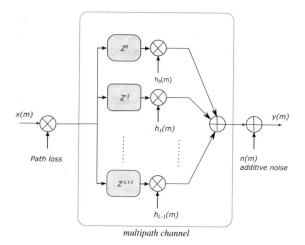

A more quantitative way to describe the effects of the multipath propagation channel is by defining the coherence bandwidth and coherence time. The first one gives an idea of what is the minimum bandwidth for which the channel can be considered flat, and the second, for how long the channel can be considered time-invariant.

Considering a multipath propagation scenario, several scaled versions of the original signal arrive at the receiver at different time instants. The span of the arriving time of the paths is called delay spread and models the dispersiveness of the wireless channel. The channel time spread is defined (conceptually) as the difference between the arrival time of the last significant replica $\tau_{\max}$, and the first arriving component τ_0, as follows

$$T_d = \tau_{\max} - \tau_0. \tag{2.17}$$

The delay spread creates frequency-selective fading. The selectivity is quantified in terms of coherence bandwidth B_c that is inversely proportional to the delay spread and is given by

$$B_c = \frac{1}{2T_d}. \tag{2.18}$$

If the signal bandwidth is smaller than the coherence bandwidth, the channel is considered flat in the frequency domain. Otherwise, when the signal bandwidth is larger than the coherence bandwidth, the channel appears as frequency selective.

A frequency-selective channel produces intersymbol interference (ISI) at the receiver that can be compensated by using equalizers. The length of the equalizer increases with the number of coefficients of the multipath channel, leading to an increment in the receiver implementation complexity. A low-complexity implementation of the equalizer can be achieved if the transmitted signal bandwidth is smaller than the coherence bandwidth of the channel, as in the OFDM case.

The physical motion of the transmitter, receiver, and reflecting objects produces variations in the channel coefficients. These variations depend on changes in the phase of the received replicas. Let us define the Doppler spread D_s as the maximum phase difference between significant paths that contribute to the coefficient l. Then, the coherence time can be defined as:

$$T_c = \frac{1}{2D_s}.$$

(2.19)

Therefore, the channel can be considered as slowly fading (stationary), when the duration of the signal pulse T_s is smaller than T_c. On the other hand, the channel results fast fading if $T_s > T_c$

2.4   Baseband Block: Multicarrier Modulation

The most popular multicarrier modulation is the so called orthogonal frequency-division multiplexing (OFDM). The idea behind OFDM is to reduce the complexity of the multipath channel equalization. To this aim, the symbol streaming is parallelized into orthogonal subcarriers with a proportional reduction in the symbol time. If it is well designed, each subcarrier is affected by an equivalent flat channel that is trivially equalized by means of a complex coefficient [11]. The complexity reduction in the equalization is an important advantage compared with that of an equivalent single-carrier system. Therefore, OFDM is the chosen technique for combating multipath fading in high data rate systems over mobile wireless channels.

An OFDM system with N subcarriers and a single antenna in the transmitter and receiver is considered is this section. A complete block diagram of an OFDM transceiver is illustrated in Fig. 2.4.

In the transmitter, the incoming symbols $X(\ell, k)$ are grouped in blocks of length N. Each of these blocks can be represented by the vector $\boldsymbol{X}(\ell) = [X(\ell, 0), X(\ell, 1), \ldots, X(\ell, N-1)]^T$, where n denotes the OFDM symbol index.

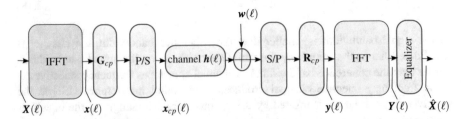

Fig. 2.4 OFDM system model

Next, OFDM modulated symbols are obtained by applying the inverse discrete Fourier transform (IDFT) to $X(\ell)$. Then, the time domain multicarrier signal $x(\ell, n)$, for $n = 0, \cdots, N - 1$, is given by

$$x(\ell, n) = \frac{1}{\sqrt{N}} \sum_{k=0}^{N-1} X(\ell, k) \exp\left(\frac{j2\pi nk}{N}\right) \quad 0 \leq n \leq (N - 1), \tag{2.20}$$

where n denotes the sub-symbol index. In order to simplify the notation, the modulation process can be represented in matrix form as

$$\mathbf{x}(\ell) = \mathbf{Q}_N \mathbf{X}(\ell), \tag{2.21}$$

where $\mathbf{x}(\ell) = [x(\ell, 0), \ldots, x(\ell, N - 1)]^T$ and the IFFT matrix has elements

$$[\mathbf{Q}_N]_{m,n} \frac{1}{\sqrt{N}} \exp\left(\frac{j2\pi mn}{N}\right) \quad \text{for } 0 \leq m, n \leq N - 1. \tag{2.22}$$

To avoid ISI and inter block interference (IBI), a guard interval is used. The most common choice is to use a cyclic prefix (CP), i.e., to copy the last L_{cp} samples to the beginning of the symbol. The cyclic prefix (CP) is appended to the original OFDM symbol and removed in the receiver avoiding both the IBI and the ISI (If L_{cp} is chosen larger than the channel length).

The transmitted OFDM symbol with CP can be expressed as:

$$\mathbf{x}_{cp}(\ell) = \mathbf{G}_{cp}\mathbf{x}(\ell) = \mathbf{G}_{cp}\mathbf{Q}_N \mathbf{X}(\ell), \tag{2.23}$$

where $\mathbf{G}_{cp}$ is an $(N + L_{cp}) \times N$ matrix that represents the cyclic prefix insertion operation and is given by

$$\mathbf{G}_{cp} = \begin{bmatrix} \mathbf{0}_{L_{cp} \times (N - L_{cp})} & \mathbf{I}_{L_{cp}} \\ \mathbf{I}_{N - L_{cp}} & \mathbf{0}_{(N - L_{cp}) \times L_{cp}} \\ \mathbf{0}_{L_{cp} \times (N - L_{cp})} & \mathbf{I}_{L_{cp}} \end{bmatrix}. \tag{2.24}$$

The sequence is transmitted through the N_h-tap wireless channel, denoted by the $N_h \times 1$ vector $\mathbf{h}(\ell)$. To simplify the analysis, channel taps are assumed constant within the OFDM block (block fading assumption) and the channel impulse response (CIR) length is considered shorter than the cyclic prefix length, i.e., $N_h < L_{cp}$. Under these assumptions, and after CP removal, the received signal can be expressed as:

$$\mathbf{y}(\ell) = \check{\mathbf{H}}(\ell)\mathbf{x}(\ell) + \mathbf{w}(\ell), \tag{2.25}$$

where $\check{\mathbf{H}}(\ell) = \mathbf{R}_{cp}\mathbf{H}(\ell)\mathbf{G}_{cp}$ is the equivalent channel matrix, $\mathbf{R}_{cp}$ is the CP removal $N \times (N + L_{cp})$ matrix, the $\mathbf{H}(n)$ represents the wireless channel of dimension $(N + L_{cp}) \times (N + L_{cp})$, and $\mathbf{w}(n)$ is the white Gaussian channel noise.

The cyclic prefix insertion and removal operation renders the linear convolution to a circulant convolution operation, i.e., $\check{\mathbf{H}}(\ell)$ is a circulant matrix. Note that this is not true for other election of prefix.

Finally, the discrete Fourier transform (DFT) operation is performed on each received block, i.e., $\mathbf{Q}_N^H \mathbf{y}(\ell)$, and the following frequency-domain symbol at subcarrier k is obtained

$$Y(\ell, k) = H(\ell, k)X(\ell, k) + W(\ell, k), \tag{2.26}$$

where $H(\ell, k)$ is the channel frequency response at subcarrier k in the time instant ℓ and $W(\ell, k)$ the additive noise assumed to be circular complex Gaussian. Equation (2.26) demonstrates the advantage of the implementation of an OFDM system. The received data symbol $Y(\ell, k)$ depends on the transmitted data $X(\ell, k)$ and channel frequency response $H(\ell, k)$. The recovering of the transmitted symbol can be done using a single-tap frequency domain equalizer (FEQ). The soft estimate of $X(\ell, k)$ can be obtained as

$$\hat{X}(\ell, k) = E_c(\ell, k)Y(\ell, k), \tag{2.27}$$

where $E_c(\ell, k)$ is the equalizer tap to be used on subcarrier k. In case of zero-forcing or *least squares* (LS) equalizer, the equalizer coefficient for each subcarrier k is given by $E_c(\ell, k) = 1/\hat{H}(\ell, k)$ where $\hat{H}(\ell, k)$ is an estimate of the channel frequency response. Alternatively, a minimum mean squared error (MMSE) equalizer can be used to take into account noise amplification at frequencies where the channel has spectral nulls [12].

2.5   Power Amplifiers: Nonlinear Distortion

Power amplifiers are one of the most important components of the analog front-end. They are responsible for a considerable portion of the transmitter power consumption and limit the system performance due to their nonlinear response.

Reduced cost mobiles handsets (mobile phones) or terminals (WLAN, WiMAX) imposes power consumption constraints that require power-efficient amplifiers. On the other hand, OFDM based systems are highly affected by the power amplifier nonlinearity. Unfortunately, there is a trade-off between power efficiency and linearity which makes difficult to achieve both in practice, at the same time.

The operation of the PA close to the saturation region is required in order to maximize its power efficiency. However, this region presents a nonlinear behavior creating nonlinear distortion (NLD) whose effect is even more noticeable when high peak-to-average power ratio (PAPR) OFDM signals are considered. The nonlinear behavior creates in-band and out-of-band distortions that degrade the performance of the system and also affect neighboring bands.

The operation point of the power amplifier is defined by its collector current (bipolar transistors) or drain current (field-effect transistors). Class A, AB, B, and C amplifiers are defined as a function of the conduction angle of the PA, which determines the performance of the amplifier in terms of linearity and power efficiency. Class A amplifiers operate with a constant current, independently of the input signal, obtaining an excellent linearity. However, the maximum power efficiency is limited to 50% when continuous wave signals are considered. This theoretical value falls significantly when the PA is excited with multicarrier signals obtaining power efficiency values lower than 5%. At the expense of linearity, class B operation improves the power efficiency reducing the conduction angle to 180°. However, the distortion introduced by class B operation (crossover) is not tolerable in most of digital communication applications. Class AB operation is an intermediate solution between class A and class B amplifiers. It increases the conduction angle to minimize the crossover distortion and reaches reasonable levels of power efficiency. Class C amplifiers present good efficiency but poor linearity and are not suitable for multicarrier digital communication systems.

2.5.1 PAPR and Power Efficiency

The OFDM signal is a sum of N complex exponential signals, as defined in (2.20). The phase and magnitude of these signals are determined by the random symbols on different subcarriers. The PAPR is defined as the ratio between the maximum instantaneous power and its average power as

$$PAPR[x(n)] = \frac{\max_{0 \leq n \leq N-1} |x(n)|^2}{E[|x(n)|^2])},$$

(2.28)

where $x(n)$ is the time domain OFDM signal sampled at symbol rate, and max{} denotes the peak value.

The PAPR of discrete-time sequences (sampled sequences) determines the complexity of the hardware required for digital signal processing, e.g., the resolution of ADC. Moreover, to avoid nonlinear distortion at the peaks, the power amplifier must be linear over a large dynamic range. The linearity and efficiency of a power amplifier are compromising specifications [13]. Consequently, the power efficiency of an OFDM system is seriously affected by its PAPR characteristics. Several approaches have been developed to obtain the distribution of the PAPR of an OFDM signal [14–16]. All these approaches assume that the OFDM signal can be approximated by a complex Gaussian random process. Under this assumption, the probability that the PAPR is below a threshold λ is given by the cumulative distribution function (CDF)

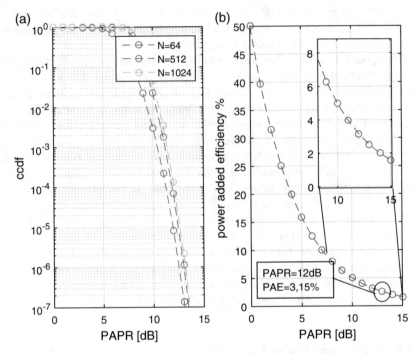

Fig. 2.5 (**a**) CCDF of PAPR distribution of an OFDM signal with $N = 64$, $N = 512$, and $N = 1024$, and (**b**) the power added efficiency (PAE) of a class A amplifier in function of PAPR

$$Pr\{PAPR[x(n)] \leq \lambda\} = [1 - \exp(-\lambda)]^N. \qquad (2.29)$$

The complementary cumulative distribution function (CCDF), $1 - Pr\{PAPR[x(n)] \leq \lambda\}$ employing (2.29) for $N = 64$, 512 and 1024 is illustrated in Fig. 2.5a.

Before transmission, the OFDM signals are amplified by a PA which is always peak-power limited. If the input signal is larger than the saturation point, the output signal will be clipped generating distortion and increasing the bit error rate and spectral regrowth. The average input power must be adjusted so that the signal is rarely clipped.

Considering a class A amplifier, the power efficiency as a function of PAPR can be written as:

$$\rho_{PA} = \frac{0.5}{PAPR}, \qquad (2.30)$$

where it can be observed that the maximum efficiency, i.e., 50% is reached when a constant amplitude signal is amplified ($PAPR = 1$).

For OFDM signals, the efficiency is dramatically reduced, as illustrated in Fig. 2.5b. A reasonable clipping probability level, for reduced system degradation, is 10^{-4} [17] obtained with a $PAPR = 12$ dB for $N = 1024$ (see Fig. 2.5a). It means that the PA is operating 12 dB above the average input power and reaches a power efficiency close to 3%. If the PAPR is reduced to 10 dB, the efficiency is increased to 6%. By replacing class A by class AB amplifiers, an efficiency of 10–15% can be obtained. However, this configuration is more nonlinear, what reduces the system performance.

These poor results in terms of efficiency motivate the development of digital signal compensation techniques, as predistortion and PAPR reduction methods, with the goal of relieving the linearity constraints and improving the power efficiency.

2.5.2   Power Amplifier Models

The models of PA can be divided into two major groups: physical models and behavioral models. Physical models require the knowledge of the electronic components of the PA, i.e., transistor parameters, and are not considered in this book.

Behavioral models (black-box models), in which no a priori knowledge of the PA's internal composition is assumed, provide good accuracy with reasonable complexity for simulation and analysis purposes. These empirical models are developed based on a set of selected input-output observations and offer a compact representation of the PA characteristics.

A brief review of the more representative black-box approaches including models with and without memory is presented in the following.

2.5.2.1   Memoryless Power Amplifier Models

If the PA output depends only of the instantaneous value of the input signal, the PA response is completely characterized by their AM/AM (amplitude to amplitude) and AM/PM (amplitude to phase) conversion characteristics. In this case, the PA has a flat frequency response over the operation bandwidth and its electro-thermal memory effects are neglected [18].

Several memoryless PA models have been proposed. These models are fitted to AM/AM and AM/PM measurements obtained using a power-swept single tone input signal. Considering the complex envelope $x(n)$ of the PA input, the complex envelope of a memoryless PA output $y(n)$ can be modeled as:

$$y(n) = g[x(n)] = F_a(|x(n)|) \exp \left\{ j \left(arg[x(n)] + F_p[|x(n)|] \right) \right\}, \qquad (2.31)$$

where $F_a(\cdot)$ and $F_p(\cdot)$ are the AM/AM and AM/PM characteristics, respectively.

Many models have been presented for a particular type of PA. Saleh model is employed for modeling high-power amplifiers such as traveling-wave tube amplifiers (TWTAs) [19]. On the other hand, when low and medium power PAs are considered, the soft limiter (SL) and the solid state power amplifier (SSPA) models are used [20].

The SL model is defined as

$$y(n) = \begin{cases} \frac{A_c}{v} x(n) & \text{for } |x(n)| \leq \frac{A_s}{A_c} v \\ A_s & \text{for } |x(n)| > \frac{A_s}{A_c} v \end{cases}, \quad (2.32)$$

where A_s is the PA saturation voltage, $A_c = A_s / \text{E}\{|x(n)|^2\}$ is the clipping level, and v^2 is the input back-off (IBO). The IBO is a scaling factor at the PA input that allows a trade-off between efficiency and linearity. Alternatively, the SSPA is defined as

$$y(n) = \frac{|x(n)|/v}{\left[1 + (|x(n)|/vA_s)^{2p}\right]^{1/2p}} \exp(j\angle\{x(n)\}), \quad (2.33)$$

where p is a model parameter that adjusts the smoothness of the transition from the linear region to the saturation region.

Polynomial models are extensively used because they are not restricted to an specific PA structure. Polynomial models give a simple mathematical expression that can be utilized for theoretical analysis and also in simulations, including the case when linearization or compensation techniques are under evaluation.

Finally, a useful model employed for theoretical evaluation is the linearized model. The Bussgang's theorem states that a Gaussian signal (valid for OFDM signals) after passing a nonlinear power amplifier $g[\cdot]$ can be written as: [21]

$$y(n) = g[x(n)] = K_L x(n) + w_d(n), \quad (2.34)$$

where $w_d(n)$ is the nonlinear distortion term uncorrelated with $x(n)$ and K_L is a scaling factor which depends on the power amplifier transfer function and its operation point.

Then, we can calculate the parameters of the equivalent linear model (2.34) as follows:

$$K_L = \frac{\text{E}\{x^*(n)g[x(n)]\}}{\text{E}\{|x(n)|^2\}}$$

$$\sigma_d^2 = \text{E}\{|g[x(n)]|^2\} - |K_L|^2 \text{E}\{|x(n)|^2\}, \quad (2.35)$$

where σ_d^2 is the variance of $w_d(n)$. The solution of (2.35) for the model in (2.32) results in

$$K_L(v) = \frac{A_c}{v}\left(1 - \exp(-v^2) + \frac{\sqrt{\pi}v}{2}\text{erfc}(v)\right) \quad (2.36)$$

$$\sigma_d^2(v) = \frac{A_c}{v^2}\left(1 - \exp(-v^2)\right) - K_L^2(v). \quad (2.37)$$

2.5.2.2   Power Amplifier Models with Memory

In broadband applications, the PA presents frequency selectivity on the operation band. As a result, the output of the PA at a given time instant depends on previous input signal values leading to a nonlinear system with memory. Several models can be employed to characterize the behavior of broadband power amplifiers. Volterra series has good modeling capacity for weakly nonlinear systems. However, they are unattractive for real-time application due to the large implementation complexity [22]. Volterra simplifications as Wiener, Hammerstein, and the Wiener–Hammerstein models also have a good modeling capacity with a reasonable complexity [23, 24].

Memory polynomial (MP) models are a good alternative for real-time applications due to their modularity and simplicity [25]. The MP model can be written as:

$$y_{MP}(n) = \sum_{m=0}^{M-1} \sum_{p=0}^{P-1} a_{mp} x(n-m) |x(n-m)|^P, \tag{2.38}$$

where a_{mp} denotes the memory polynomial coefficients, P and M are the polynomial order and memory depth, respectively.

Other polynomial models can be found in the literature. For example, the orthogonal memory polynomial [26] that can alleviate the instability problem associated with the conventional polynomial models, and the generalized memory polynomial [27] which combines the memory polynomial with additional cross terms improving estimation algorithm stability at a small increase of the computational complexity.

2.6   Low Noise Amplifiers

The most important figure of merit of an LNA is its noise performance. The LNA is the first device placed after the reception antenna and must provide enough gain without adding additional noise. High gain is also a mandatory requirement for a LNA in order to obtain an entire receiver chain with low noise figure, as described by (2.9).

For a LNA selection, the most important characteristics such as gain, noise figure, linearity, and power consumption are considered in a figure of merit, that it is expressed by Rappaport et al. [28]

$$FoM_{LNA} = \frac{G_{LNA} IIP3 f}{(NF - 1) P_{LNA}}, \tag{2.39}$$

where NF is the noise figure (linear scale), G_{LNA} is the LNA gain, $IIP3$ is the third-order input intersection point and P_{LNA} is the consumed power.

As discussed in previous sections, the receiver architecture and modulation type define the requirements of the LNA.

2.7   Mixers: Phase and Amplitude Imbalances

The RF front-end is composed of critical non-ideal components such as digital-to-analog converters, low-pass filters, the IQ modulator, and the power amplifier which seriously affect the system performance.

At the transmitter, the digital baseband OFDM signal $x(n)$ is converted to a continuous-time (complex) baseband signal, $x(t)$, with in-phase (I) and quadrature (Q) components $x_i(t)$ and $x_q(t)$. These signals are filtered by low-pass filters, $h_i(t)$ and $h_q(t)$, that model the cascade of DACs and the analog filters employed to eliminate Nyquist images and noise (impulse responses $h_i(t)$ and $h_q(t)$ are in general different, creating frequency-dependent IQ mismatch).

In-phase and quadrature components at the low-pass filters' output (continuous-time baseband signals) are directly modulated to RF using two signals with an offset of 90° and a frequency f_c, from the local oscillator (LO). Considering practical implementations, LO signals present phase and amplitude imbalance between the IQ branches [29]. A block diagram of IQ imbalance model is illustrated in Fig. 2.6.

The continuous-time passband signal can be written as:

$$\tilde{x}_{rf}(t) = (x_i(t) \otimes h_i(t))c_i(t) - (x_q(t) \otimes h_q(t))c_q(t), \qquad (2.40)$$

where $\otimes$ denotes convolution.

Denoting β and θ as the amplitude imbalance and the phase orthogonality mismatch, respectively, of the IQ modulator [29], the in-phase and quadrature carrier signals including IQ mismatch, are given by

$$c_i(t) = \cos(2\pi f_c t), \quad c_q(t) = \beta \sin(2\pi f_c t + \theta)). \qquad (2.41)$$

By replacing (2.41) in (2.40) and defining

$$g_1(t) = \frac{1}{2}\left(h_i(t) + \beta e^{j\theta} h_q(t)\right) \quad g_2(t) = \frac{1}{2}\left(h_i(t) - \beta e^{j\theta} h_q(t)\right), \qquad (2.42)$$

Fig. 2.6
Frequency-dependent IQ imbalance model

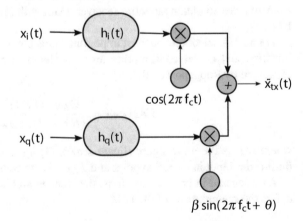

the equivalent signal at the output of the modulator can be expressed as:

$$\tilde{x}_{tx}(t) = \left(g_1(t) \otimes x(t) + g_2(t) \otimes x^*(t)\right). \tag{2.43}$$

At the receiver side, the signal is downconverted, filtered, and sampled to recover the original symbols. Assuming an ideal demodulator, the signal is sampled at period T, that gives $x_{tx}(n) = x_{tx}(nT)$. Under this assumption, the discrete-time baseband output signal can be written as:

$$\tilde{x}(n) = g_1(n) \otimes x(n) + g_2(n) \otimes x^*(n), \tag{2.44}$$

and the equivalent frequency-domain baseband signal at subcarrier k is given by

$$X_{tx}(k) = \left(G_1(k)X(k) + G_2(k)X^{\#}(k)\right), \tag{2.45}$$

where $X^{\#}(k) = X^*(N - k)$ represents the mirrored OFDM symbol [30] obtained from the frequency domain representation of $\mathbf{x}(n) = [x(n, 0), x(n, 1), \ldots, x(n, N-1)]^T$ and $\mathbf{x}^*(n) = [x^*(n, 0), x^*(n, 1), \ldots, x^*(n, N-1)]^T$ vectors, and $G_1(k)$ and $G_2(k)$ are the frequency response of the equivalent filters $g_1(n)$ and $g_2(n)$, respectively.

To assess the impact of the IQ imbalance over the system, the image rejection ratio (IRR) is employed. IRR is the power ratio between the desired signal and the image signal (generated by the mirror component). It can be expressed for a frequency-dependent IQ imbalance as

$$IRR(k) = \frac{|G_1(k)|^2}{|G_2(k)|^2}. \tag{2.46}$$

The IRR for an IQ mixer with amplitude imbalance of 2% and phase imbalance of 2° is illustrated in Fig. 2.7, considering a mismatch between the equivalent low-pass filters h_i and h_q. The difference is quantified by the impulse response mismatch $b(m)$ defined as $h_i(m) = b(m) \otimes h_i(m)$. It can be observed that the IRR is frequency independent when $b(m)$ is a Dirac impulse ($h_i(m) = h_q(m)$). The parameters of the IQ modulator are $\beta = 2\%$ and $\theta = 2°$, with several filter mismatches: (a) $b(z) = 1 + 0.05z^{-1}$, (b) $b(z) = 1 + 0.05z^{-1}$, and (c) $b(z) = 1$.

Nonlinear distortion is also generated by active mixers, where 1 dB compression point and IIP3 are defined in similar way to that of power amplifiers. Input signal level at RF (downconverter) and baseband (up-converter) need to be adjusted in order to avoid mixer saturation.

I/Q imbalances appear also at the receiver side in the down-conversion process. Similar interference models are employed to evaluate their effects over the system performance. More details about IQ imbalances and its compensation at the receiver side can be found in [31].

Fig. 2.7 Image rejection
ratio (IRR) of IQ mixer with
amplitude and phase
mismatches $\beta = 2\%$ and
$\theta = 2°$, respectively,
including several filter
mismatches: (a)
$b(z) = 1 + 0.05z^{-1}$, (b)
$b(z) = 1 + 0.05z^{-1}$, and (c)
$b(z) = 1$

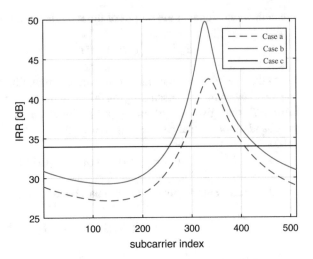

Modern communication systems as full-duplex (FD) transceivers are also affected by IQ imbalances. The implementation of a self-interference canceller, that is mandatory in FD applications, needs to consider the IQ imbalances in order to reach a reasonable performance [32, 33].

2.8   Local Oscillator: Phase Noise

Local oscillator phase noise creates two sources of distortion, namely the common phase error term (CPE) and the intercarrier interference (ICI) term. The CPE acts on all subcarriers, and causes an identical attenuation/rotation for all constellation points. It can be easily corrected by using pilots [34]. In addition, the CPE term can be modeled as a random variable with non-zero mean [35, 36].

The ICI term induced by phase noise is generated by $N-1$ symbols from adjacent subcarriers and has different value for each subcarrier making its cancellation very difficult [37, 38]. Considering that all subcarriers are affected by a random phase component, the ICI term can follow a Gaussian distribution with zero mean. The most important contribution to ICI comes from the nearest neighbor subcarriers and decreases gradually for more distant subcarriers. This is the reason why central subcarriers are the most affected by the ICI because they have large number of neighbors subcarriers. That situation corresponds to the worst case for ICI analysis.

The contribution of the phase noise (PN) to each component is a function of the phase noise bandwidth and the OFDM subcarrier spacing $\Delta f = 1/(NT_s)$. For large subcarrier spacing, the PN is concentrated around the subcarrier and the leakage over neighboring subcarriers is negligible given a large value of CPE and a reduced ICI. On the other hand, when the phase noise bandwidth is larger than the subcarrier spacing, the phase noise spreads over several subcarriers creating a large level of ICI [35, 36, 39].

To obtain the CPE and ICI contributions, we need to define a model for the local oscillator (LO). A straightforward way to characterize a LO is by its power spectral density (PSD), typically described by the following piece-wise linear function [40]:

$$S(f) = \begin{cases} 10^{-a} & 0 \le f < f_L \\ 10^{-a} \left(\frac{f}{f_L}\right)^{-b} & f_L \le f < f_p \\ 10^{-c} & f > f_p \end{cases},$$

where the parameter c defines the white phase noise floor and f_L and f_p define the phase locked loop (PLL) bandwidth and the phase noise bandwidth (region where the noise floor becomes dominant), respectively. a gives the phase noise level near to the center frequency and b gives the noise fall-off rate. These parameters are specified in Fig. 2.8a. This model is useful to approximate local oscillators based on a PLL often used in practice [41]. Generally, the LO signal quality is specified by the *integrated single-side band phase noise* (IPN), given by $IPN = \int_0^\infty S(f)df = \frac{1}{2}\sigma_{pn}^2$ (in dBc) [42]. Moreover, the *RMS integrated phase error* I_{rms}, $I_{rms} = (180/\pi)\sqrt{2\int_0^\infty S(f)df}$, specified in degrees can also be employed as a measurement of LO quality.

An example of the PSD of a local oscillator implemented with a PLL is shown in Fig. 2.8b where the oscillator PN has an IPN of -32 dBc with a loop bandwidth of $f_L = 1$ kHz and an error floor of -130 dBc. The low-frequency region of the PSD contributes to the CPE term. On the other hand, the high frequency region is the source of ICI. The cutoff frequency of the PLL loop filter is designed smaller than the intercarrier spacing. In that case, the phase noise is concentrated in the CPE component and can be easily compensated.

A figure of merit is also defined for local oscillators [43]. The FoM_{vco} is a simple but yet fair way of comparing different oscillator architectures, and can be expressed by

$$FoM_{vco} = -PN(\delta f) + 20\log\left(\frac{f_c}{\delta f}\right) - 10\log(P_{lo}), \qquad (2.47)$$

where $PN(\delta f)$ is the phase noise at offset of δf, f_c is the carrier frequency, and P_{lo} is the consumed power by the LO expressed in mW. The FoM is useful to highlight the trade-off between oscillator quality and power consumption. An adequate numerology can be made to obtain an OFDM signal robust to phase noise. As discussed, the subcarrier spacing can be selected to minimize the ICI, which is hard to compensate.

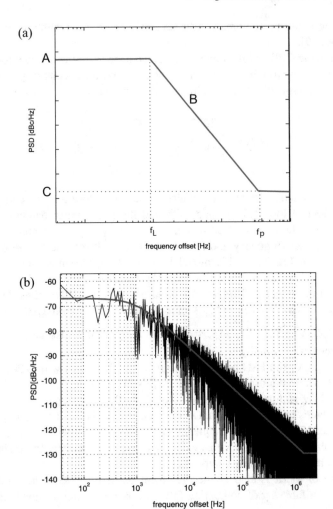

Fig. 2.8 (**a**) PSD of phase noise definition as piece-wise linear function, (**b**) PSD of the phase noise with an integrated phase noise power (IPNP) of −32 dBc, a loop bandwidth of 1000 Hz and an error floor of −130 dBc

2.8.1 Millimeter-Wave Phase Noise Modeling

Low-phase noise oscillators required for applications operating at millimeter wave bands may be expensive and power consuming for massive market devices. It is a challenging issue to consider the use of the spectrum near and above 30 GHz. The phase noise increases 6 dB per frequency doubling, reducing the effective signal-to-noise ratio and limiting the use of modulations with high constellation size [44]. A parametric phase noise model for millimetric waves is proposed in [45], where the phase noise degradation with the increment of oscillation frequency is included.

The power spectral density of the phase noise can be represented by a multi-pole/zero model given by

$$PSD(f_o) = PSDO \frac{\prod_{n=1}^{N} 1 + \left(\frac{f_o}{f_{zn}}\right)^{\alpha_{zn}}}{\prod_{m=1}^{M} 1 + \left(\frac{f_c}{f_{pn}}\right)^{\alpha_{pn}}}, \tag{2.48}$$

where the parameters can be found in [45] (tables 1, 3, and 4). The PSD of three oscillators operating at 29.55, 45, and 70 GHz are illustrated in Fig. 2.9.

To illustrate the performance degradation due to phase noise and its dependence with subcarrier spacing, an OFDM system with a subcarrier spacing varying from 10 kHz to 10 MHz is simulated. The local oscillator is implemented with a crystal oscillator and voltage-controlled oscillator (VCO) connected in a PLL. The achievable SINR subject to ICI is illustrated in Fig. 2.10. These results

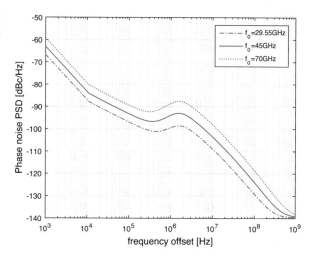

Fig. 2.9 Phase noise PSD of an oscillator with frequencies 29.55, 45, and 70 GHz

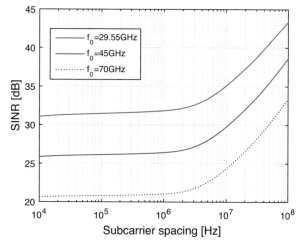

Fig. 2.10 Achievable SINR subject to intercarrier interference for an OFDM system with LO frequencies of 29.55, 45, and 70 GHz and subcarrier spacing varying from 10 kHz to 10 MHz

verify that phase noise imposes a severe constraint for the development of mm-wave frequency systems. Particularly, in OFDM implementations the number of subcarriers, bandwidth, and subcarrier spacing need to be carefully selected [46, 47].

2.9 Analog-to-Digital Converters (ADC)

An analog-to-digital converter (ADC) is a dedicated circuit design capable of sampling an analog signal at discrete time instants (determined and triggered by a clock), and translates the signal amplitude at those instants into discrete levels. Combined, these two tasks allow for sampling of the analog signal and conversion to the digital domain. However, there exist different circuit topologies that can be used to perform such conversion. In general, the selection of a particular ADC architecture is dependent on the requirements of the system to be sampled. Each ADC architecture has its own advantages and limitations when compared to the others, and there is usually a trade-off between sampling speed and resolution difficult to overcome [48].

2.9.1 ADC: Performance Metrics

Performance metrics are important when the performance of an ADC is not satisfactory in terms of conversion errors as they allow to *measure* the quality of a converter and *quantify* the effects of non-ideal behavior that degrade performance. In this sense, they provide key knowledge on what effects need to be compensated for, which performance parameters are expected to improve when doing so, and which is the maximum achievable (ideal) performance.

Transfer function linearity metrics such as integral nonlinearity (INL) and differential nonlinearity (DNL) are performance metrics that can be found in any commercial ADC datasheet and are used to give an idea of the maximal deviation of the ADC from the ideal transfer characteristic as well as a quick lower bound estimate on the loss in effective resolution for the sampling bandwidth.

However, when carefully evaluating the ADC performance, other dynamic effects that also contribute to distortion and loss in effective resolution must be considered. To this purpose, *metrics of spectral purity* such as signal-to-noise and distortion ratio (SINAD) and spurious free dynamic range (SFDR) are used, which can be directly measured from input-output data and better quantify the joint effect of all non-ideal behavior and nonlinear distortion that degrade performance.

2.9.2 Metrics of Spectral Purity

In this subsection are defined and described general performance metrics that allow to assess the spectral purity of an ADC and therefore the amount of distortion present at its output:

Signal to quantization noise ratio (SQNR) is the ratio between the input signal power (P_S) at full scale and the power of output quantization noise (P_{QN}), which is the difference between the analog input signal and its ideally distortion free quantized version. This defines the maximum dynamic range of an ideal ADC and is usually calculated by applying a sinusoid input signal. In this case, for $b \geq 4$ the quantization noise can be considered white with uniform distribution in the range $[-\Delta/2, \Delta/2]$ and it can be shown that

$$SQNR[dB] = \frac{P_S}{P_{QN}} \cong 6b + 1.72, \tag{2.49}$$

where b is the ADC resolution in number of bits.

Signal to noise and distortion ratio (SINAD) is the ratio between the input signal power (P_S) at full scale and the power of output quantization noise plus the power of distortion components at the quantized output signal ($P_{QN} + P_D$),

$$SINAD[dB] = \frac{P_S}{P_{QN} + P_D}. \tag{2.50}$$

Effective number of bits (ENOB) is the equivalent resolution in a real ADC affected by non-ideal behavior and distortion, and can be computed from the SINAD expression in (2.50) using the right-hand side of (2.49) as

$$ENOB = (SINAD - 1.72)/6. \tag{2.51}$$

Spurious free dynamic range (SFDR) is the difference between the power of a sinusoid input signal measured at the output of the ADC and the power of the highest harmonic distortion tone.

2.9.3 Transfer Function Linearity Metrics

Integral Nonlinearity (INL) is the difference between the analog voltage V_k that should trigger a digital output code transition in an ideal ADC (say from code $k - 1$ to k), and the voltage $V[k]$ that actually causes that transition in the real ADC under consideration. INL is computed after correcting gain (G) and offset (V_{off}) by minimizing

$$\Upsilon[k] = V_k - GV[k] - V_{off} \tag{2.52}$$

using measurements at each possible code transition k, and expressed as a percentage of the ADC digital range as

$$INL[k] = 100 \frac{\Upsilon[k]}{2^b \Delta}, \tag{2.53}$$

where Δ is the quantization step, i.e., the analog voltage equivalent to a least significant bit (LSB) and b is the ADC nominal resolution. According to [48], the reduction in effective resolution due to INL can be accurately modeled by

$$b_{loss} = 0.5 \log_2 \left(1 + 3INL^2\right). \tag{2.54}$$

Differential Nonlinearity (DINL) is the difference $\omega[k] = INL[k+1] - INL[k]$ between the analog input voltages that cause two consecutive code transitions in the real ADC under test, when compared with the ideal value Δ. It can be defined in relation to Δ as:

$$DNL[K] = \frac{\omega[k] - \Delta}{\Delta}. \tag{2.55}$$

Figure 2.11 shows the transfer characteristic of an ideal ADC (dashed line) and that of a real ADC (solid line) where examples of INL and DNL are shown.

Fig. 2.11 INL and DNL

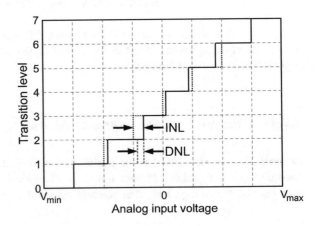

2.10   Digital-to-Analog Converters (DAC)

A digital-to-analog converter (DAC) is the circuit used to translate a digital signal back to the analog domain. In the following subsections we present a brief description of the most commonly used available architectures [48, 49].

2.10.1   Binary Weighed DAC

Figure 2.12 shows a current driven binary weighed switching digital-to-analog converter (DAC). In this topology, each bit in the input digital signal is used to turn on (or off) a particular switch, allowing a current proportional to the bit value to add (or not) to the current I_o in the load (resistance) in order to get the analog output voltage. A disadvantage of this implementation as shown in Fig. 2.12 is that this direct current switching produces glitches at the output, but there are available techniques such as current steering that can significantly alleviate this effect [49]. However, the main concern on this architecture is that due to mismatches, transitions from several least significant bits turned on to a most significant bit (e.g., from "0111" to "1000") can lead to considerable errors in the output analog voltage level. Because this particular type of transition involves switching several current sources off and turning only one on, should there be mismatch in the components, the sum of the least significant currents might be lower than expected when compared to the most significant one or even higher, producing an important amount of nonlinear distortion at the analog output. In the following section we will briefly describe an alternative architecture that prevents this effects from occurring.

Fig. 2.12 Binary weighed DAC

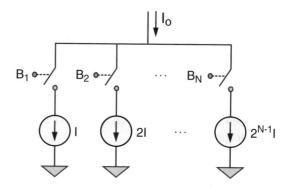

Fig. 2.13 Segmented DAC

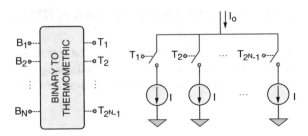

2.10.2 Segmented DAC

Another approach is depicted in Fig. 2.13, where a binary to thermometric coding is performed prior to the current switching stage. In this case, $2^b - 1$ equal valued current sources are used instead, and therefore each transition from one binary code to the next (or previous) involves switching on (or off) only one current source. As a result, the transfer characteristic is guaranteed to be monotonically increasing and the errors due to mismatch are minimized, as they do not add up during transitions between two consecutive codes.

However, this solution can dramatically increase the amount of switches (or current steering circuitry) if the required DAC resolution is high ($b \geq 10$ bits). Thus, as an alternative *partial segmentation* can be used. That is, use binary weighing for the least significant bits where the errors due to mismatch can be kept low, and segmentation for the most significant bits.

2.11 Summary of the Key Points

- Several analog front-end impairments affect severally the performance of the communication system and they must be considered in the link budget.
- Linearity and power efficiency are two conflicting demands for power amplifiers. The situation is even worse when high-PAPR signals like OFDM are considered.
- Phase noise from local oscillators destroys the OFDM orthogonality when high carrier frequency and/or reduced subcarrier spacing are evaluated. The OFDM numerology must be chosen adequately in order to mitigate the harmful effects of the phase noise.

References

1. M. Shafi, A.F. Molisch, P.J. Smith, T. Haustein, P. Zhu, P. De Silva, F. Tufvesson, A. Benjebbour, G. Wunder, 5G: a tutorial overview of standards, trials, challenges, deployment, and practice. IEEE J. Sel. Areas Commun. **35**(6), 1201–1221 (2017)
2. F. Gregorio, *Analysis and Compensation of Nonlinear Power Amplifier Effects in Multi-Antenna OFDM Systems*, Ph.D. thesis (Helsinki University of Technology, Helsinki, 2007)

3. F. Gregorio, J. Cousseau, S. Werner, T. Riihonen, R. Wichman, EVM analysis for broadband OFDM direct-conversion transmitters. IEEE Trans. Veh. Technol. **62**(7), 3443–3451 (2013)
4. C. Zhao, R.J. Baxley, Error vector magnitude analysis for OFDM systems, in *Signals, Systems and Computers, (ASILOMAR), 2006* (2006), pp. 1830–1834
5. R. Gomes, L. Sismeiro, C. Ribeiro, T.R. Fernandes, M.G. Sanchéz, A. Hammoudeh, R.F.S. Caldeirinha, Will COTS RF front-ends really cope with 5G requirements at mmWave? IEEE Access **6**, 38745–38769 (2018)
6. A. Mammela, A. Anttonen, Why will computing power need particular attention in future wireless devices? IEEE Circuits Syst. Mag. **17**(1), 12–26 (2017)
7. A. Molisch, *Wireless Communications*, 2nd edn. (Wiley, Chichester, 2011)
8. J. Wood, *Behavioral Modeling and Linearization of Power Amplifiers* (Artech House, London, 2014)
9. ETSI, LTE; evolved universal terrestrial radio access (E-UTRA); base station (BS) conformance testing, in *3GPP TS 36.141 version 12.6.0 Release 1* (2015)
10. D. Tse, P. Viswanath, *Fundamentals of Wireless Communication* (Cambridge University Press, Cambridge, 2005)
11. L. Hanzo, W. Webb, T. Keller, *Single- and Multi-carrier Quadrature Amplitude Modulation* (Wiley, Chichester, 2000)
12. A. Pandharipande, Principles of OFDM. IEEE Potentials **21**(12), 16–19 (2002)
13. J. Vuolevi, T. Rahkonen, *Distortion in RF Power Amplifiers* (Artech House, Norwood, 2003)
14. R. Prasad, *OFDM for Wireless Communications Systems* (Artech House, Boston, 2004)
15. H. Ochiai, H. Imai, On the distribution of the peak-to-average power ratio in OFDM signals. IEEE Trans. Commun. **49**(2), 282–289 (2001)
16. R. Van Nee, A. De Wild, Reducing the peak-to-average power ratio of OFDM, in *Proceeding IEEE Vehicular Technology Conference, VTC'98*, vol 3 (IEEE, Ottawa, 1998), pp. 2072–2076
17. R.J. Baxley, G.T. Zhou, Power savings analysis of peak-to-average power ratio in OFDM. IEEE Trans. Consum. Electron. **50**(3), 792–798 (2004)
18. S. Boumaiza, F.M. Ghannouchi, Thermal memory effects modeling and compensation in RF power amplifiers and predistortion linearizers. IEEE Trans. Microwave Theory Tech. **51**(12), 2427–2433 (2003)
19. A.A.M. Saleh, Frequency-independent and frequency-dependent nonlinear models of TWT amplifiers. IEEE Trans. Commun. **29**(11), 1715–1720 (1981)
20. S. Cripps, *Advanced Techniques in RF Power Amplifiers Design* (Artech, Boston, 2002)
21. J.J. Bussgang, Cross correlation function of amplitude-distorted Gaussian input signals, in *Research Laboratory of Electronics, Massachusetts Institute of Technology, Cambridge, MA, Technical Report 216*, vol 3 (1952)
22. M. Schetzen, *The Volterra and Wiener Theories of Nonlinear Systems* (Wiley, Hoboken, 1980)
23. E. Aschbacher, M. Rupp, Modeling and identification of a nonlinear power-amplifier with memory for nonlinear digital adaptive pre-distortion, in *Proceeding IEEE Signal Processing Advances in Wireless Communications, SPAWC 2003*, vol 1 (IEEE, Rome, 2003), pp. 658–662
24. P. Crama, J. Schoukens, Initial estimates of Wiener and Hammerstein systems using multisine excitation. IEEE Trans. Instrum. Meas. **50**(6), 1791–1795 (2001)
25. O. Hammi, F. Ghannouchi, B. Vassilakis, A compact envelope memory polynomial for RF transmitters modeling with application to baseband and RF digital pre distortion. IEEE Microwave Wireless Compon. Lett. **18**(5) (2008)
26. R. Raich, H. Qian, G.T. Zhou, Orthogonal polynomials for power amplifier modeling and predistorter design. IEEE Trans. Veh. Technol. **53**(5), 1468–1479 (2004)
27. D. Morgan, Z. Ma, J. Kim, M. Zierdt, J. Pastalan, A generalized memory polynomial model for digital predistortion of RF power amplifiers. IEEE Trans. Signal Process. **54**(10), 3852–3860 (2006)
28. T. Rappaport, R.W. Heath Jr., R.C. Daniels, J.N. Murdock, *Millimeter Wave Wireless Communications* (Pearson/Prentice Hall, New Jersey, 2014)
29. C.-L. Liu, Impacts of I/Q imbalance on QPSK-OFDM-QAM detection. IEEE Trans. Consum. Electron. **44**(3), 984–989 (1998)

30. A. Tarighat, R. Bagheri, A.H. Sayed, Compensation schemes and performance analysis of IQ imbalances in OFDM receivers. IEEE Trans. Signal Process. **53**(8), 3257–3268 (2005)
31. M.A. Ali, M. Arif, W. Kumar, Joint CIR, CFO, DCO and FI/FS Rx IQ imbalance estimation. IET Commun. **10**(15), 2025–2033 (2016)
32. J. Wang, H. Yu, Y. Wu, F. Shu, J. Wang, R. Chen, J. Li, Pilot optimization and power allocation for OFDM-Based full-duplex relay networks with IQ-Imbalances. IEEE Access **5**, 24344–24352 (2017)
33. D. Korpi, L. Anttila, V. Syrjälä, M. Valkama, Widely linear digital self-interference cancellation in direct-conversion full-duplex transceiver. IEEE J. Sel. Areas Commun. **32**(9), 1674–1687 (2014)
34. S. Stefanatos, F. Foukalas, T. Khattab, On the achievable rates of OFDM with common phase error compensation in phase noise channels. IEEE Trans. Commun. **65**(8), 3509–3521 (2017)
35. C. Muschallik, Influence of RF oscillators on an OFDM signal. IEEE Trans. Consum. Electron. **41**(3), 592–603 (1995)
36. L. Piazzo, P. Mandarini, Analysis of phase noise effects in OFDM modems. IEEE Trans. Commun. **50**(10), 1696–1705 (2003)
37. G. Liu, W. Zhu, Phase noise effects and mitigation in OFDM systems over Rayleigh fading channels. Wireless Pers. Commun. Springer **41**(2), 243–258 (2007)
38. A. Leshem, M. Yemini, Phase noise compensation for OFDM systems. IEEE Trans. Signal Process. **65**(21), 5675–5686 (2017)
39. J. Stott, The effects of phase noise in COFDM, in *BBC Research Development Technical Review, Summer* (1998)
40. K. Fazel, S. Kaiser, *Multi-Carrier and Spread Spectrum Systems. From OFDM and MC-CDMA to LTE and WiMAX*, 2nd edn. (Wiley, Hoboken , 2008)
41. J. Tubbax, B. Come, L. Van der Perre, S. Donnay, M. Engels, H. De Man, M. Moonen, Compensation of IQ imbalance and phase noise in OFDM systems. IEEE Trans. Wireless Commun. **4**(3), 872–877 (2005)
42. J.L. Stensby, *Phase-Locked Loops. Theory and Applications* (CRC Press, Boca Raton, 1997)
43. L. Fanori, P. Andreani, Highly efficient class-C CMOS VCOS, including a comparison with class-B VCOS. IEEE J. Solid-State Circuits **48**(7), 1730–1740 (2013)
44. S. Andersson, S. Mattisson, Design considerations for 5G mm-wave receivers, in *2017 Fifth International Workshop on Cloud Technologies and Energy Efficiency in Mobile Communication Networks (CLEEN)* (2017), pp. 1–5.
45. Ericsson, Further elaboration on mm-wave phase noise modelling, in *3GPP TSG-RAN WG4, R4-1703087* (2017)
46. J. Vihriälä, A.A. Zaidi, V. Venkatasubramanian, N. He, E. Tiirola, J. Medbo, E. Lähetkangas, K. Werner, K. Pajukoski, A. Cedergren, R. Baldemair, Numerology and frame structure for 5G radio access, in *2016 IEEE 27th Annual International Symposium on Personal, Indoor, and Mobile Radio Communications (PIMRC)* (2016), pp. 1–5
47. A.A. Zaidi, R. Baldemair, H. Tullberg, H. Bjorkegren, L. Sundstrom, J. Medbo, C. Kilinc, I. Da Silva, Waveform and numerology to support 5G services and requirements. IEEE Commun. Mag. **54**(11), 90–98 (2016)
48. R. Van de Plassche, *CMOS Integrated Analog-to-Digital and Digital-to-Analog Converters* (Kluwer Academic, Dordrecht, 2003)
49. B. Razavi, The current-steering DAC [A circuit for all seasons]. *IEEE Solid-State Circuits Magazine* **10**(1), 11–15 (2018)

Chapter 3
Energy Consumption

Abstract It is generally assumed that the analog front-end (AFE) dominates the power consumption of the system. However, in low-power applications the digital block can dominate the consumption of the transceiver. Typical scenarios in the upcoming 5G technologies include ultra-dense small-cell networks, where the distance between the base-station and users is small (or even inexistent), such that low power transmission is sufficient. In this context, the digital signal processing power consumption dominates over the analog power dissipation. In order to handle these new scenarios, in addition to the power dissipated in the AFE it is necessary to take into account the power consumption in the baseband processing. For example, the trade-off between the energy consumed to implement channel coding and the energy saved due to code gain needs to be carefully evaluated. Sometimes, the implementation of coding is not useful in terms of energy efficiency. Digital signal processing and AFE power consumption and scaling of the power consumption of the different devices are addressed in this chapter, considering typical system parameters as transmitted power, operation bandwidth, and modulation size. Energy efficiency (EE) and spectral efficiency (SE) are used to assess the system performance.

3.1 Introduction

Nowadays, energy consumption is one of the main concerns in the design and operation performance of wireless communication systems. For this reason, energy efficiency has been chosen as a new figure of merit to be optimized. Novel communication systems as 5G will have to consider energy efficiency as one of its strong performance points. Spectral efficiency, data rate, throughput, latency, and device cost are also significant issues to take under consideration [1].

© Springer Nature Switzerland AG 2020
F. Gregorio et al., *Signal Processing Techniques for Power Efficient Wireless Communication Systems*, Signals and Communication Technology,
https://doi.org/10.1007/978-3-030-32437-7_3

Communication systems support a massive number of devices demanding high data rates, ubiquitous access, and uninterrupted coverage. This exponential growth in the system capacity raised the expansion of network infrastructures and an escalation of energy demand. Hence, the main challenge is to satisfy the high data traffic increasing the energy efficiency of the wireless networks.

Several approaches are developed to meet the requirements of high data traffic, keeping the energy consumption in a moderate level. These techniques can be categorized as follows [2, 3].

- Resource allocation: the system allocates the power budget in the most suitable bands to maximize the energy efficiency, allowing a moderate reduction in the SE.
- Network planning: conventionally the networks are deployed in order to maximize the covered region. However, if the network is deployed maximizing the covered area per unit of consumed energy, a significant improvement in terms of energy efficiency can be reached [4, 5].
- Energy harvesting: to reduce the consumption of non-green energy, communication systems can harvest renewable energy from the environment. Typical sources are the sun and the wind. Novel alternatives include radio frequency signals or mechanic vibrations [6, 7].
- Hardware optimization: hardware is designed to operate in a high energy-efficient region. In mid to long range communications, the RF front-end is the major energy consumer of the system. The energy efficiency can be improved by carefully designing the transmitted waveform, selecting the appropriate ADC and DAC resolution, and adjusting the power amplifier operation point. Digital power consumption also need to be considered, particularly in short range applications.

In this chapter, the last approach is addressed. The power consumption (PC) of each element is studied, and its dependence with the operation bandwidth, the number of antennas, the transmission power, and the data rate are described. The RF imperfections and their effects on the energy efficiency of the communication system are also addressed [8]. To make the analysis more realistic, the consumption of the DSP-based RF impairment compensation techniques is included in this analysis [9].

3.2   Energy Efficiency and Spectral Efficiency

The limitations of available spectrum and the demand of a huge data rate motivate the development of high spectral efficiency (SE) techniques. The adoption of multicarrier modulation techniques such as OFDM dramatically improved the system data rate increasing the spectral efficiency. However, the energy efficiency (EE) has not been enhanced accordingly. Furthermore, linearity requirements imposed to the power amplifier response generate power inefficient solutions.

The energy efficiency is defined as the channel capacity normalized by the system power consumption. The total PC can be expressed by

$$P_s = \frac{P_t}{\rho} + P_c,$$

(3.1)

where ρ is the power amplifier efficiency, P_t is the transmitted power, and P_c is the electronic circuit power consumption. The circuit PC can be separated in several modules as the base-band (BB) module, the radio-frequency (RF) module, the power amplifier (PA) module, and the active cooling and power supply modules [10] as is illustrated in Fig. 3.1. Considering that the channel capacity is defined by the Shannon's formula

$$C = W \log_2 \left(1 + \frac{P_t}{W N_0} \right),$$

(3.2)

where W is the system bandwidth and N_0 stands for the power spectral density of the additive white Gaussian noise, and the spectral efficiency (SE) is given by

$$\eta_{SE} = \frac{C}{W} = \log_2 \left(1 + \frac{P_t}{W N_0} \right).$$

(3.3)

To evaluate the trade-off between energy efficiency and spectral efficiency is given by the quotient between the spectral efficiency and the consumed power

$$\eta_{EE} = \frac{\eta_{SE}}{\left(\frac{P_t}{\rho} + P_c \right)}.$$

(3.4)

The energy efficiency is a function of the transmitted power, the power amplifier efficiency, and the electronic circuit power consumption.

In systems with large transmission power, as macrocell cellular base stations, the power amplifier dominates the total power drained. In addition, multicarrier modulated signals such as OFDM present high levels of PAPR. As a result, PAs must operate at higher IBO due to the high linearity requirements, lowering its power efficiency and increasing the power expended. On the other hand, in low-power systems as femtocells, power consumption is dominated by baseband processing.

In this scenario, the signal processing algorithms need to be optimized in order to keep the used power at moderate levels.

The distribution of the power consumption in typical base stations, from macrocell BS to femtocell BS, is illustrated in Fig. 3.2. It can be categorized in two groups:

- Fixed power consumption: the drained power is independent of the transmitted power. This group includes power consumption of baseband processing and RF AFE.

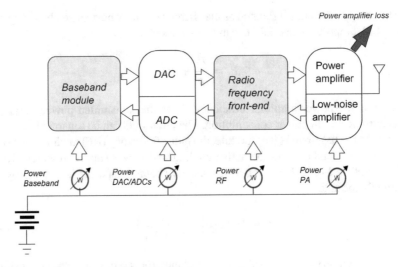

Fig. 3.1 Power consumption of each block

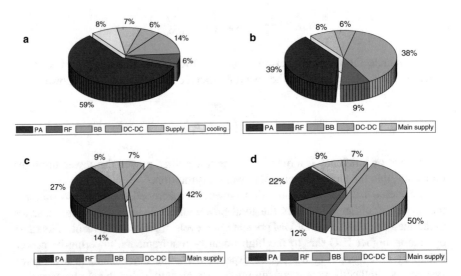

Fig. 3.2 Contribution of each block to the total power consumption of several base stations. (**a**) Macrocell base station, (**b**) microcell base station, (**c**) picocell base station, and (**d**) femtocell base station

- Power transmission dependent consumption: Power amplifier expended scales with the transmitted power. This leads to additional dissipation due to losses in DC-DC power supply, mains supply, and active cooling, which scale linearly with the consumed power. Their effects can be approximated by the efficiency factors ρ_{DC}, ρ_{MS}, and ρ_{cool}, respectively.

The consumption of a single transmitter chain can then be expressed by

$$P_c = \frac{\frac{P_t}{\rho} + P_{RF} + P_{BB}}{\rho_{DC}\rho_{MS}\rho_{cool}} \tag{3.5}$$

From this equation it is possible to infer that when the transmission power is increased, the PA is the main power consumer, and the relative influence of the fixed consumers (RF and BB blocks) is reduced. In this scenario, the efficiency of the PA is the main concern. Signal processing algorithms, as PAPR reduction techniques and predistortion methods, are well justified. On the other hand, when a low power transmission is used, like in pico and femtocells base stations, the baseband processing becomes significant [11]. In this case, the use of linearization methods and PAPR reduction techniques could be reconsidered. The power savings due to an improvement in PA efficiency provided by the these techniques could be less than the power required to implement them.

3.3 Digital Block and Front-End Power Consumption Models

A communication system is composed of multiple transceivers (TX/RX) each connected to one or more transmit/receive antenna elements. The transmitter includes an RF front-end, a power amplifier (PA), a digital-to-analog interface (DAC), and a digital processing block. On the other hand, the RX also includes a DSP block and an RF front-end, a low noise amplifier, and an analog-to-digital interface (ADC). The power consumption of these components usually depends on several system parameters as: modulation type, operation bandwidth, link attenuation, noise power, bit resolution, and coding/decoding methods.

In the following, the power consumption of the more significant components will be addressed, including RF and digital blocks.

3.3.1 Radio Frequency Front-End

The main components of the RF front-end are the power amplifier, synthesizers, low noise amplifiers, and mixers. The power expended by the RF block of the transmitter can be expressed by

$$P_{RF-TX} = P_{PA} + P_{syn} + P_{mix} + P_{DAC}, \tag{3.6}$$

where P_{PA}, P_{syn}, P_{mix}, and P_{DAC} denote the power consumption of the power amplifier, synthesizer, mixer, and DAC, respectively.

At the receiver, the power consumption of the RF front-end is given by

$$P_{RF-RX} = P_{LNA} + P_{syn} + P_{mix} + P_{ADC}, \tag{3.7}$$

where P_{LNA} and P_{ADC} denote the power expended by the low noise amplifier and the ADC, respectively.

The power consumption of the main components of the RF front-end are described in the following.

1. Power amplifiers (PA)

   The power consumed by a PA depends on the amplifier characteristics, dynamic range of the modulated signal, required linearity, operation bandwidth, biasing mode, and operation back-off. The direct application of an input back-off (IBO) to handle the large dynamic range and linearity requirements of the OFDM signal and achieve a low-distortion level leads to a poor power efficiency. To increase the efficiency while maintaining satisfactory levels of distortion, compensation techniques as predistortion can be used. These techniques improve the PA linearity and reduce the required IBO.

   The power consumed can be expressed by

$$P_{PA} = \frac{P_t}{\rho}, \tag{3.8}$$

   where P_t is the radiated power and ρ is the average power efficiency of the PA.

   Base stations using conventional A/AB class amplifiers reach power added efficiency (PAE) levels of around 5–10% [12]. Envelope tracking (ET) and Doherty amplifiers are able to increase the efficiency up to 60%, for LTE base stations [13, 14]. Power added efficiency levels varying from 30% to 65% are reported for several ET implementations [15–20]. Doherty amplifiers with PAE larger than 60%, for macro and microcells base stations (medium and high power), are presented in [21–23]. The PAE for several PAs types, considering commercial and prototype implementations, are illustrated in Fig. 3.3.

2. Synthesizer

   The local oscillator signal is generated by a phase locked loop (PLL) synthesizer, coupled with a voltage controlled oscillator (VCO). A figure of merit for local oscillators FoM_{vco}, evidencing the trade-off between oscillator quality (phase noise) and power consumption, is defined in (2.47).

   VCOs are typically controlled through a synthesizer loop, which adds a considerable complexity to the front-end. The power consumption of the frequency synthesizer is composed by the combination of VCO and PLL power consumption, i.e., $P_{syn} = P_{vco} + P_{pll}$.

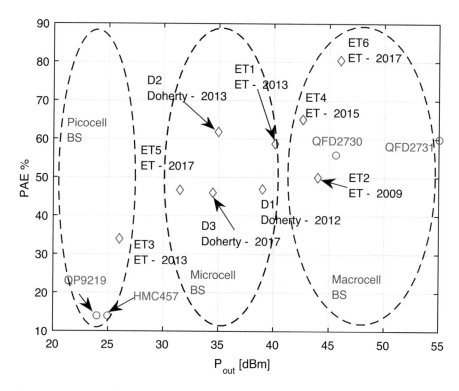

Fig. 3.3 Power amplifiers: power added efficiency (PAE) vs transmitted power for several architectures considering low, medium, and high PAs

3. Mixers

Up- and down-converters are, respectively, used in the transmitter and the receiver. The operation bandwidth, noise figure, linearity, and conversion gain are parameters that affect the mixer consumption. In order to evaluate different mixer designs, the following figure of merit was defined in [24]

$$FoM_{mix} = \frac{I_{P3}G_{mix}}{NF_{mix}P_{mix}}, \tag{3.9}$$

where I_{P3}, G_{mix}, NF_{mix}, and P_{mix} are the third-order intersection point, conversion gain, noise figure, and power expended by the mixer, respectively.

4. Low noise amplifiers (LNAs)

The main requirement of an LNA is to provide a high gain with low noise figure. The good behavior in terms of gain, noise figure, and linearity is reached at expenses of a large power consumption. This trade-off is quantified by a figure of merit FoM_{LNA}, defined in (2.39).

Table 3.1 Power
consumption of RF
components of a macrocell
BS transmitter [25]

Component	Power consumption
Mixer (IQ modulator)	925 mW
VGA	575 mW
VCO + Synthesizer	297 mW
VCO + Synthesizer (feedback)	297 mW
Feedback mixer	1000 mW
Clock	825 mW
Dual DAC	141.9 mW
ADC	504 mW
Consumed power P_{RF}	5.72 W
Expected consumed power P_{RF} (2020)	1 W

Table 3.2 Power
consumption of RF
components of a macrocell
BS receiver [25]

Component	Power consumption
LNA 1	297 mW $\times$ 2
LNA 2	850 mW $\times$ 2
Dual Downconverter Mixer	1750 mW
VGA	1200 mW
VCO + Synthesizer	297 mW
Clock	825 mW
ADC	936 mW
Consumed power P_{RF} (2012)	7.3 W
Expected consumed power P_{RF} (2020)	1.29 W

Table 3.1 summarizes the power expended by a macrocell base station transmitter
[25]. The transmitter is designed for high power systems, where digital predistortion
(PD) is mandatory to reduce the nonlinear distortion, while keeping a reasonable
PA efficiency. The table includes the power consumption of the feedback branch
required for PD implementation, composed by an ADC, a feedback mixer, and a
synthesizer. In macrocell base stations, the PA concentrates the power consumption
of the system, what motivates the optimization of the PA operation point and the
development of novel PA architectures.

In small cells, the transmitted power is low and the consumption RF front-
end and baseband processing blocks becomes significant, which motivates the
development of advanced techniques to improve the efficiency of these blocks.

The consumption of the RF components of the receiver front-end, for a similar
macrocell BS, is presented in Table 3.2. The receiver is composed by a cascade of
two LNAs (LNA 1 and LNA 2), a dual downconverter, a variable gain amplifier, and
ADC. The clock and local oscillator are shared with the transmitter block.

The reported values in Tables 3.1 and 3.2 were published in 2012. Over the years,
the consumption of the RF components decreases due to the evolution of silicon
technology and novel circuits architectures.

This evolution roughly follows a Moore's-law-like tendency, with an average
reduction of $\sqrt{2}$ in the consumed power every 2 years, when a new generation of

Table 3.3 Power consumption of RF components of a massive MIMO transmitter and receiver front-ends [27] (year 2013)

Component	Power consumption (TX)	Power consumption (RX)
Predriver	115 mW	0
Synthesizer	125 mW	125 mW
Clock	75 mW	75 mW
DAC	225 mW	0
LNA	0	25 mW
Mixer	200 mW	200 mW
VGA	0	63 mW
ADC	0	175 mW
Total P_{RF}	0.74 W	0.66 W

PA is released [26]. Using this rule, the final figures in Tables 3.1 and 3.2 can be updated to 2020 dividing by a factor $4\sqrt{2}$. This results in a consumption of 1 W for the transmitter front-end and 1.29 W for the receiver.

It should be noted that these front-ends were developed for macro BS where highly linear (and more power hungry) components are employed. For macro BS the constraints in terms of implementation costs and power consumption are relaxed compared to those of mobile devices. Components with high FoM and without high restrictions in the consumed power are employed in this scenario generating low levels of distortion (TX-side) and low values of noise figures (RX-side).

For the case of massive products, as cellular phones and pico and femto BSs, cost reasons impose the use of moderate/low quality RF components. The same situation happens in massive MIMO implementations, where it is necessary to keep low the cost and power consumption of the RF front-end. The reference power consumption for the analog blocks of a large scale MIMO system are 0.74 W and 0.76 W for the receiver and transmitter, respectively. The consumption of each component is shown in Table 3.3 [27].

LTE Cellphone Power Consumption Model

For the case of cellular phones, an empirical power consumption model is presented in [28]. The diagram of an LTE power consumption model is depicted in Fig. 3.4. It describes the consumption when the phone is active. The consumption of an user equipment (UE) can be calculated as

$$P_{UE} = P_{on} + \zeta_{DL}(P_{DL} + P_{RX,RF}(P_{RX}) + P_{RX,BB}(R_{DL}))$$

$$+\zeta_{UL}(P_{UL} + P_{TX,RF}(P_{TX}) + P_{TX,BB}(R_{UL})), \qquad (3.10)$$

where $P_{RX,RF}$ and $P_{TX,RF}$ are the power consumption of receiver and transmitter front-ends, ζ_{DL} and ζ_{UL} are logical variables that indicates the operation mode (reception/transmission), and R_{UL} and R_{DL} are the uplink and downlink average rate, respectively. P_{on} is the power consumed when the cellular subsystem is active

Fig. 3.4 LTE cellular phone power consumption model

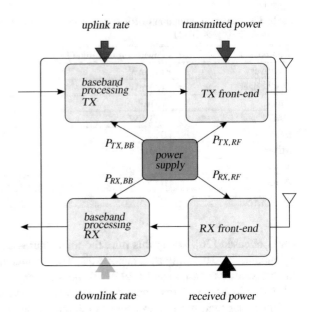

uplink rate   transmitted power

downlink rate   received power

Table 3.4 Power consumption parameters for a cellphone [28] (polynomial fit in mW)

Term	
Receiver front-end	
$P_{RX,RF}(P_{RX}) = -0.04P_{RX} + 24.8$	$P_{RX} \leq -52.5\,\text{dBm}$
$P_{RX,RF}(P_{RX}) = -0.11P_{RX} + 7.86$	$P_{RX} > -52.5\,\text{dBm}$
Transmitter front-end	
$P_{TX,RF}(P_{TX}) = 0.78P_{TX} + 23.6$	$P_{TX} \leq 0.2\,\text{dBm}$
$P_{TX,RF}(P_{TX}) = 17.0P_{TX} + 45.4$	$0.2\,\text{dBm} < P_{TX} \leq 11.4\,\text{dBm}$
$P_{TX,RF}(P_{TX}) = 5.90P_{TX}^2 - 118P_{TX} + 1195$	$11.4\,\text{dBm} < P_{TX}$
Baseband processing	
$P_{RX,BB} = 0.62$	
$P_{TX,BB} = 0.97R_{DL} + 8.16$	
Cellular subsystem	
P_{on}	853
P_{UL}	29.9
P_{DL}	25.1

(display, CPU), and P_{DL} and P_{UL} are the power consumed when the receiver and the transmitter units are actively receiving and transmitting information, respectively. $P_{RX,BB}$ and $P_{TX,BB}$ denote the power consumption of baseband processing at the receiver and the transmitter, respectively.

The variable terms for the empirical model fitted to an LTE Release 8 category 3 cellphone are defined in Table 3.4 [28]. This model shows the dependence of power consumption with transmitted and received power and the uplink/downlink average rate. The dependence of the consumption with the operation bandwidth is not contemplated in this model. Power consumption and its dependence with operation bandwidth and also the carrier aggregation operation mode are addressed in [29].

To reduce the overall power consumption, the phone can operate in idle and discontinuous reception (DRX) sleep modes. These operation modes are not addressed in this chapter.

3.3.2 Baseband Processing

The power expended by the digital block depends on the number of arithmetic operations required to implement different tasks. The consumption depends on the employed technology, the hardware, and the required resolution [30].

Baseband block executes the FFT/IFFT for OFDM, modulation/demodulation, signal synchronization, channel estimation, equalization, and channel coding/decoding. Additionally, large power systems require to carry out the digital predistortion, PAPR reduction techniques, and compensation of RF non-idealities.

The metric for power efficiency in baseband processing is the ratio of the number of operations per second to the power consumed to execute such operations. It is expressed in Giga complex arithmetic operations per second per Watt (GOPS/W). For dedicated implementations, considering 4 quantization bits, an efficiency of 400 GOPS/W was reported in [26] in 2015. The efficiency increases with the evolution of the technology, achieving an increment of 20% every 2 years [25]. However, the digital power consumption is composed by the sum of static and dynamic power. The static power is caused by leakage currents that diminishes the increment in the efficiency when migrating to dense technologies.

A block diagram of the main tasks to be executed in a OFDM transceiver is shown in Fig. 3.5. The transmission/reception process is described in the following [31]

- Coding:

  Channel coding includes redundant information to allow the error correction and provide better error rate performance compared with the uncoded system. The error correction capacity is a function of the amount of added redundancy. The information to be transmitted is encoded at rate $r_{cod} = \frac{k_0}{s_0}$, where k_0 and s_0 are the number of bits at the input and output of the encoder, respectively. The complexity of channel encoder is considered low and negligible when compared with the implementation complexity of the decoding process.

- Mapping:

  The bits s_0 are grouped and mapped to a constellation of symbols of size M_p. The symbol rate is given by $R_s = \frac{R}{r_{cod} \log_2(M_p)}$, where R is the input rate.

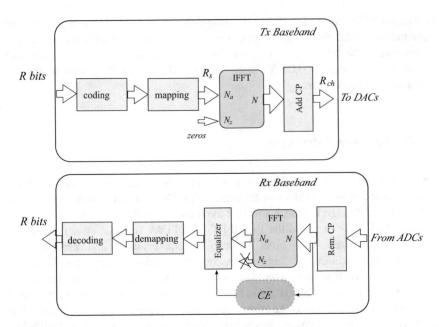

Fig. 3.5 Block diagram of baseband processing required for an OFDM transceiver implementation

- OFDM modulation:

  A modulation of block length N is implemented via an IDFT of the same size. The N subcarriers are composed by N_a active subcarriers and N_z zero subcarriers. After the IDFT, a cyclic prefix of length L_{cp} is added. Then, the symbols stream is converted into the analog domain using DAC and feed to the front-end. The DAC conversion rate is equal to $R_{ch} = R_s N(1 + L_{cp})/N_a$.

  Considering an FFT split-radix algorithm, the implementation of an N-point IDFT/DFT requires P_{FFT} real multiplications and A_{FFT} real additions

  $$P_{FFT} = N(\log_2(N) - 3) + 4$$
  $$A_{FFT} = 3N(\log_2(N) - 1) + 4. \tag{3.11}$$

  If each active subcarrier transmits b bits, the normalized data bit per symbol is expressed by $D_i = N_a b/(1 + L_{cp})$.

- OFDM demodulation:

  After downconversion and analog-to-digital conversion, the cyclic prefix is removed, and the symbols are demodulated using a DFT. OFDM modulation/demodulation process requires identical implementation complexity.

- Equalizer:

  A single-tap frequency domain equalizer is implemented to compensate for the channel response. Considering N_a subcarriers, the equalizer implementation requires $P_{eq} = 4N_a$ real multiplications and $A_{eq} = 2N_a$ additions. This calculation assumes that the frequency domain channel response is available at the receiver side. However, additional operations are needed for channel estimation and the calculation of the equalizer taps.

- Demapping and decoding:

  The received complex symbols need to be demapped and decoded. The channel decoder usually has a significant computational cost. As an example, a turbo code based on the log-MAP algorithm decoder is considered next. The total number of additions per information bit of the iterative decoder is given by

$$A_{dec} = \frac{2I\left(25.2^{v} + 13\right)}{k_0},\tag{3.12}$$

where I is the number of iterations, k_0 are the output bits, and v is the code memory.

As it is noted from (3.12), the complexity of the decoding increases with the number of iterations. Therefore, the trade-off between the error correction capacity and the implementation complexity needs to be evaluated. Error control codes are used to increase the link reliability and reduce the transmitted power. However, an extra computational cost is required to implement the decoder at the receiver.

The power consumption of the decoding process needs to be compared with the coding gain and energy savings at the transmitter. An energy-efficient code must verify that the power saving due to the code gain is larger than the energy required to implement the decoder. This trade-off is determined by the link distance and the transmitted power. For short-distance links, low-complexity coders can be implemented to obtain energy-efficient solutions.

The selection of the coding techniques depends on the scenario and the wireless devices [32]. In applications with limited energy budget, only low-complexity implementation can be considered. Hamming, Reed–Solomon, and convolutional codes are the preferred options allowing reasonable performance with moderate implementation complexity. On the other side, systems without power restrictions at the receiver and operating over long distances, make use of complex codes to reduce the transmitted power and improve the link reliability. In this scenario, turbo and LDPC codes are the best options.

An example of the trade-off between code gain and implementation complexity is presented in Table 3.5. The table shows LDPC, turbo (TCC), Hamming, and

Table 3.5 Coding gain and decoding complexity of LDPC, turbo, Hamming, and convolutional codes with code rate of 1/2, a sequence of bits of $N = 1452$, and $E_b/N_o = 1.6\,\text{dB}$ in AWGN channel [33]

Code	Operations/iter	Iterations	Total operations	Coding gain
TCC	437	5.1	2228	8.5
LDPC	76	32	2432	8.3
Hamming	336	1	336	3
Convolutional	217	1	217	4.8

convolutional codes along with their main characteristics: coding gain and number of required operations. These results consider a code rate of $1/2$, a sequence of 1452 bits, and $E_b/N_o = 1.6$ dB in AWGN channel [33].

3.4 ADC: Power Consumption, Resolution, and Sampling Frequency Trade-Off

High-performance ADCs are in general power hungry devices. In addition, the higher the required performance in terms of resolution and (particularly) sampling speed, the higher the power consumption. This is due to the fact that power consumption in ADCs is directly related to the circuit complexity of the particular architecture implementation. For instance, an N bit flash ADC, which provides the highest sampling rate among single stage converters, uses 2^N comparators. Thus an increment of just one bit in the ADC resolution means the circuit complexity and power consumption is doubled. This is one of the reasons why this architecture is limited in practice to low or medium resolution (normally between 4 and 6 bits). However, resolution can be increased by means of a pipelined structure, in which several ADCs are connected in a cascaded manner such that a coarse quantization is obtained in the first ADC and finer resolution is obtained in subsequent stages. This is a popular conversion strategy as it allows for a linear increment in circuit complexity as a function of resolution. Modern wideband communications systems may require even higher conversion speed in order to comply with the required sampling rates. In this case, sampling speed can be further increased with a linear increment in circuit complexity by using a time-interleaved structure, where M ADCs are connected in parallel branches with shifted clocks such that the sampling rate is increased by M.

On the other hand, algorithmic ADCs can provide high resolution with reduced circuit complexity and thus power consumption, as is the case of successive approximation register (SAR) ADCs or sigma-delta converters. However, the conversion speed is limited in the first case by the required N clock cycles to complete conversion for an N-bit output word, and by the required oversampling ratio (higher than 8) in the second. Therefore, there is a trade-off between resolution, sampling speed, and power (defined by circuit complexity) that must be carefully addressed when choosing a sampling strategy for a particular application.

As a rule of thumb, the highest achievable performance in terms of sampling speed for a given resolution can be obtained by time-interleaved pipelined flash ADCs at the cost of very high circuit complexity and power consumption, whereas similar resolution can be obtained with the lowest power consumption by SAR or sigma-delta ADCs but at limited conversion rates (see Chap. 5 for references and more details on ADC architectures). In the following subsections, a more detailed discussion is presented.

3.4.1 ADC Figures of Merit and Approximate Power Consumption

Recently, there has been a significant progress in lowering the required conversion energy in ADCs, regardless of the particular architectures and following the trade-off between resolution, conversion speed, and consumption [1]. In the case of ADCs, FoMs are kept as simple as possible by focusing on effective resolution, conversion rate, and power dissipation. While one of the first used FoM that adjusts well for high-speed ADCs assumed a power dissipation doubled by each extra bit in the ENOB, recent studies suggest that for medium to high resolution ADCs the dissipated power is actually multiplied by four for each extra bit, which is more consistent with noise-limited analog circuits. In [34], the suggested FoM follows the latter case and is given by

$$FoM = SINAD[dB] + 10 \log \left(\frac{f_s/2}{P_c} \right), \qquad (3.13)$$

where $SINAD$ represents the achievable dynamic range taking into account the associated distortion of a practical implementation, P_c is the consumed power, and $f_s/2$ represents the conversion bandwidth. In most cases, the actual conversion bandwidth is lower than $f_s/2$ due to the use of a slight oversampling, which turns the FoM into an upper bound regarding this parameter.

The power consumption (PC) of ADCs, particularly when high performance in terms of conversion rate and resolution is required, can have a significant impact on the overall consumption in modern wideband receivers. As previously discussed, ADCs PC is in general related to the architecture chosen to meet the sampling requirements of the system. However, an approximate expression that gives an idea on how the PC of ADCs changes as a function of resolution and conversion rate is introduced in [35], and given by

$$P_c = C \left(2^{2ENOB} \right) f_s^{\upsilon}, \qquad (3.14)$$

where C is a constant representing the PC of a reference ADC, and υ is 1 for noise-limited ADCs and 2 for technology limited high-speed ADCs. Equation (3.14) illustrates the quadratic relation between the PC and the sampling frequency as well as the exponential increment as a function of effective resolution in recent ADC designs where $f_s > 1\,GHz$.

3.4.2 Lower Bound on Power Consumption for Noise-Limited ADCs

A detailed derivation of lower bounds for the power consumption on noise-limited ADCs is presented in [36]. These values are calculated based on lower bounds for the PC of different circuits that are usually used as components of most ADCs, and have been found to be around 100 times lower than the PC of state-of-the-art ADCs at the time of publication.

The first component analyzed is the sample and hold circuit, present at the input of most ADCs. In this case, the lower bound on the PC can be computed as

$$P_S = 24kT \left(2^{2b}\right) f_s \tag{3.15}$$

where k is the Boltzmann constant, T is the temperature in Kelvin, and b is the ideal ADC resolution. Note that this *noise-limited* approximation underestimates the growth in PC as a function of the sampling frequency, taking its influence as a linear relation rather than the more realistic quadratic term. This is because the quadratic term is a better fit for *technology-limited, very high-speed* ADCs. However, (3.15) is still useful as a lower bound and serves as a base to analyze other components.

The analysis in [36] continues with the PC of other two fundamental components of ADCs, which are the comparators and amplifiers. For the case of the comparators, we have

$$P_C \cong b \ln(2) \frac{V_{eff}}{V_{FS}} P_S \approx 2P_S, \tag{3.16}$$

where V_{eff} is the voltage needed to drive the current used to charge the load capacitor in the required settling time, and V_{FS} is the full scale voltage at the ADC input. In the case of power amplifiers, the following bound can be derived:

$$P_A \cong \left(1 + (1 + G)b \ln(2) \frac{V_{eff}}{V_{FS}}\right) \frac{1 + G}{G^2} P_S \approx 3P_S, \tag{3.17}$$

where G is the gain of the power amplifier. Using these components as reference, it is possible to estimate the power consumption of some ADC architectures. For example, in a pipelined structure, the power dissipation in an efficient design scales down on a factor of two per stage, where the main consumption is that of the inter-stage power amplifiers. Then

$$P_P \cong P_A \sum_{i=0}^{b-2} 2^{-i} \approx 2P_A. \tag{3.18}$$

We can carry out a similar analysis for flash converters, where the number of comparators grow exponentially with resolution dominating the PC, which can thus be approximated by

$$P_F \cong P_S + (2^b - 1) P_C. \tag{3.19}$$

Note that these are very optimistic lower bounds, but their behavior is in accordance with the brief analysis introduced at the beginning of this section, and they offer a guide for estimating the PC in several ADCs. The actual numeric results can be adjusted for each design if a precise estimation of the PC of their components is known.

3.5 Short-Range and Long-Range Links

For short-range communication systems where energy-constrained devices are employed, the required processing power is similar than the transmit power [37–39]. In this case, the impact of the power consumption of analog and DSP blocks can be significant.

Channel coding is implemented to reduce the transmitted power while keeping a suitable error rate. However, the computational cost of coding implementation needs to be considered in short-distance systems, as discussed in Sect. 3.3.2. Having in mind this code gain-implementation consumption trade-off, it is evident that there exists a link distance for which the total power consumption of a system with and without coding is the same, for identical BER. This threshold is denominated as the crossover distance for coding. It results then than for values below this distance, the power consumption required for coding implementation is larger than the transmitted power reduction obtained by using coding.

In this section, a study of power efficiency for a short-range communication system is presented. In the analysis, a simplified transceiver model is employed considering only the more significant components. On the receiver side, the LNA, ADC, and baseband processing are considered. On the transmitter, the emphasis is put in the power amplifier and baseband processing, since according to the discussion in Sect. 3.3 they are the main source of power consumption.

Considering a wireless link with two nodes placed at a distance d, and assuming free-space propagation, the received power can be expressed by

$$P_{rx} = \frac{\lambda^2}{4\pi} d^n P_t, \tag{3.20}$$

where P_t is the transmitted power, n is the pathloss exponent, and λ is the wavelength of the carrier signal. By defining the signal-to-noise ratio as

$$\frac{S}{N} = \frac{RE_b}{BN_0} = \eta_{SE}\frac{E_b}{N_0}, \tag{3.21}$$

where η_{SE} is the spectral efficiency, B is the bandwidth, R is the system rate, E_b is the energy per bit, and N_0 is the noise. The system noise N can be written as a function of the operation bandwidth and thermal noise as $N = mkT_eB$, where m is a proportionality constant that includes the receiver noise figure, k is the Boltzmann constant, and T_e is the absolute temperature.

Combining the previous equations, the minimum transmitted power to achieve a desired signal-to-noise ratio can be written as:

$$P_t = \frac{S}{N}N\left(\frac{4\pi}{\lambda^2}\right)^2 = \eta_{SE}\frac{E_b}{N_0}mkTB\left(\frac{4\pi}{\lambda^2}\right)^2. \tag{3.22}$$

Considering a wireless link without channel coding, the minimum transmitted power to reach a desired error rate is given by

$$P_{t_U} = \eta_{SE_U}kT_eB10^{(SNR_U/10)+NF/10)}\left(\frac{4\pi}{\lambda^2}\right)^2 d^n, \tag{3.23}$$

where $SNR_U = E_b/N_o$ is the uncoded SNR required for an specified BER, NF is the receiver noise figure, and η_{SE_U} is the uncoded spectral efficiency.

To reduce the transmission power, channel coding is implemented.

The coded system can obtain the same BER than the implementation without coding with lower SNR. The difference in the required SNR to obtain identical BER is known as the coding gain, $CG = SNR_U - SNR_C$. There is a trade-off between the coding gain and its implementation complexity. Long codes provide high coding gain, but with a higher power consumption. This compromise is evaluated in the following.

The minimum required transmitted power using channel coding can be expressed in a similar way than (3.23) as follows:

$$P_{t_C} = \eta_{SE_c}KTB_{cc}10^{(SNR_c/10)+NF/10)}\left(\frac{4\pi}{\lambda^2}\right)^2 d^n$$

$$= \frac{P_{t_U}}{10^{(SNR_U-SNR_c)/10}} = \frac{P_{t_U}}{10^{CG/10}}, \tag{3.24}$$

where $\eta_{SE_c} = R/B_{cc} = R_c$ is spectral efficiency of a coded system. R_c denotes the coding rate that is a function of the added redundancy.

The reduction of the transmitted power can be quantified by the energy saving per information bit that is expressed by

$$E_{sav} = \frac{P_{t_U} - P_{t_C}}{R} = \frac{P_{t_U}}{R}\left(1 - 10^{-CG/10}\right). \tag{3.25}$$

Table 3.6 Characteristics of RS and LDPC codes: coding gain, maximum rate, and consumption

Decoder	Coding gain (CG)	P_{max} (mW)	R_{max} (Mbps)	V_{dd}	E_{dec}
RS (255,239)	2	58	160	1.8	0.1193
LDPC $N=1024$	6.1	630	500	1.5	0.56

The energy saving obtained with the implementation of channel coding must include the energy required to execute the decoding process. The effective energy saving can be written as:

$$\tilde{E}_{sav} = \frac{P_{tx_U}}{R} \left(1 - 10^{-CG/10}\right) - E_{dec}. \tag{3.26}$$

The decoding power consumption is composed by two components: the dynamic and the static power consumption. The dynamic power consumption is associated with the switching capacity and depends on the operation frequency and the power supply voltage. Leakage currents and biasing sources are responsible for the static component. Considering a large operation frequency, i.e., large throughput, the dynamic component is the dominant and the static part can be neglected.

Therefore, the energy per decoded bit can be calculated by using

$$E_{dec} = \frac{P_{max}}{R_{max} V_{dd}^2}, \tag{3.27}$$

where P_{max} is the maximum power consumption of the decoding process at the maximum rate R_{max}, and V_{dd} is the power supply voltage. These values depend on the implemented coded and the employed process size. The decoder parameters are shown in Table 3.6.

The crossover distance, when the coding gain is compensated by the decoding consumption, can be found from (3.26) under the condition $\tilde{E}_{sav} = 0$. Then, the distance results

$$d_{th} = \left(\frac{\lambda^2 E_{dec}}{(4\pi)^2 kT_e 10^{(SNR_U/10)+NF/10)} \left(1 - 10^{-CG/10}\right)}\right)^{1/n}. \tag{3.28}$$

The crossover distance using LDPC and RS codes, considering pathloss exponents of $n = 2, 3, 4$, and a frequency operation range from 100 MHz to 10 GHz are illustrated in Fig. 3.6. The curves are obtained using a temperature of $T = 300$ K, a receiver noise figure of $NF = 5$, and uncoded SNR of 15 dB.

In the previous basic analysis, only the thermal noise at the receiver is considered. However, in a more realistic scenario, the impairments introduced by the power amplifier and the ADC have a strong influence in the equivalent transmitted SNR. After including this imperfections, the required transmitted power results

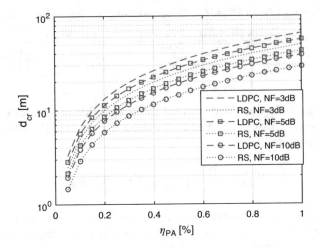

Fig. 3.6 Crossover distance in function of power amplifier efficiency and receiver noise figure considering LDPC and RS codes

$$P_{t_U} = (1 + \rho)\eta_{SE_U}kT_eB10^{(S\tilde{N}R_U/10)+NF/10)}\left(\frac{4\pi}{\lambda^2}\right)^2 d^n, \qquad (3.29)$$

where ρ denotes the power efficiency of the PA, operating in the linear region. The quantization noise of the ADC of b_{adc} bits is included in the effective SNR as $S\tilde{N}R_U = SNR_U + 10\log_{10}(P_q)$, where $P_q = 2^{-2b_{adc}}$. It must be noted that $S\tilde{N}R_U$ decreases for ADC with lower resolution. The drop in the SNR must be compensated by increasing the transmitted power.

This trade-off needs to be evaluated in the selection of the ADC resolution. The dependence of the crossover distance with the power amplifier efficiency and receiver noise figure is illustrated in Fig. 3.7.

Transmission and computational power consumption due to the coding process are evaluated for several data rates: $R = 10$ Mbps, 100 Mbps, and 1 Gbps, using LDPC and RS codes. In this example, the computational power P_{cc} is associated with the power required in the decoding process. The transmission power is the sum of the transmitted power and the PA consumption. The intersection point between P_{cc} and P_t determines the crossover distance where the implementation of the coding technique becomes energy efficient. For distances between nodes larger than the crossover distance, the implementation of a codification technique saves energy compared with the uncoded system. These results are illustrated in Figs. 3.8 and 3.9. The simulations consider an ideal ADC converter, a power amplifier efficiency $\rho = 0.25$, and an uncoded SNR of 15 dB. From these results, it can be inferred that the selection of coding techniques is determined by the transmission power and the distance between nodes. The use of codes like LDPC, that provide large coding gain with high implementation complexity, is justified for link distances larger than

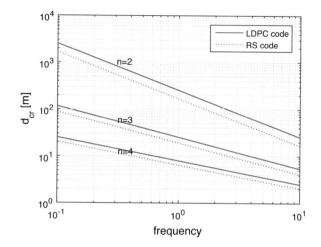

Fig. 3.7 Crossover distance in function of channel pathloss exponent considering LDPC and RS codes

Fig. 3.8 Power consumption of transmitter and decoder using RS codes. A pathloss exponent $n = 3$ and PA efficiency $\rho = 0.25$ are employed. An ideal ADC is considered

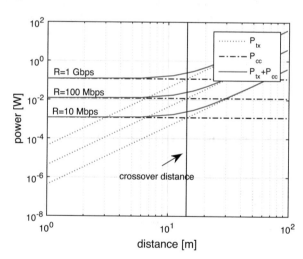

30 m. On the other hand, RS codes, that provide lower coding gain but better of energy efficiency, are preferable for shorter links.

3.6 Power Consumption Scaling

In order to reach large data rate, the use of large constellation size is one of the main options. Moreover, the increment of operation bandwidth and the use of large number of antennas can be also considered to obtain high communication rates.

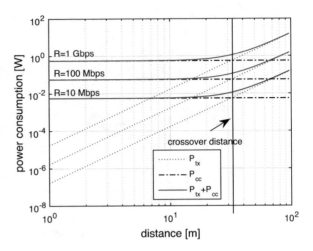

Fig. 3.9 Power consumption of transmitter and decoder using LDPC codes. A pathloss exponent $n = 3$ and PA efficiency $\eta_{PA} = 0.25$ are employed. An ideal ADC is considered

However, the modification of these parameters impact on the power consumption of the transceiver. Operation bandwidth, constellation size, and number of antennas are the main parameters to be considered in terms of system performance and power consumption [40].

In the following, these parameters and their effects over the power consumption are described.

3.6.1   Bandwidth Dependence

The increment of the operation bandwidth requires the use of ADCs with higher sampling frequency. As is expressed in Sect. 3.4, the power consumption of the ADC scales linearly with the sampling frequency and exponentially with bit resolution. The power consumption can be expressed as

$$P_c = \frac{FoM}{f_s 2^{ENOB}},$$
(3.30)

where FoM is the figure of merit that depends on the architecture and the implementation technology of the ADC [34, 41].

The power consumption is a bottleneck when multiantenna systems with large bandwidth are considered. Several works are focused on the reduction of the ADC power consumption by reducing the needed bit resolution [42, 43]. Particularly, in massive MIMO implementations, the quantization noise effects are minimized due to the robustness of massive MIMO systems to hardware impairments [40]. This topic is addressed with more details in the Chap. 8.

3.6.2   Number of Antennas

In MIMO systems, digital beamforming techniques are implemented to steers the signals generated from a set of transmit antennas to a predetermined angular direction. In massive MIMO cases, the number of front-ends scales linearly with the number of antennas, increasing the cost and power consumption. However, the massive MIMO scenario allows the use of RF components of low cost, moderating the power consumption. It is verified that massive MIMO systems support larger distortion levels, while maintaining a high throughput [44, 45].

The implementation of pure baseband digital beamforming requires a separate RF chain per antenna that results in large power consumption and high implementation cost. It motivates the use of hybrid beamforming that operates in the baseband and analog domains.

Hybrid beamforming provides a cost-saving solution by reducing the number of front-ends, with a moderate performance degradation compared to conventionally fully-digital implementations [46, 47].

3.6.3   Data Rate (Constellation Size)

Modulation with large constellation size and robust coding are required to increase the system data rate. Dense constellations impose severe constraints in the distortion generated by the RF components. There are several components that affect the quality of the signal. At the transmitter side, I/Q imbalance in the mixer, phase noise in local oscillator, and power amplifier nonlinear distortion are the main contributions. On the other hand, distortion in the LNA, phase noise in local oscillator, I/Q imbalances, and ADC quantization noise affect the receiver side.

The quality of the transmitted signal is quantified by the error vector magnitude (EVM). Standards define an allowed EVM level that is function of the constellation size and the code rate. For example, the allowed levels specified for the downlink in the 3GPP standard are -21.9 dB for 64 QAM constellation, -18 dB for 16 QAM, and -15 dB for QPSK [48]. One of the main contributors to the signal distortion is the PA. The PA operation point needs to be set in a region such that the EVM requirements are satisfied. A large back-off is required to meet the specified EVM for large constellation size, increasing the PA power consumption. Additionally, the effect in the EVM of the phase noise, I/Q imbalance, and ADC resolution also need to be considered.

At the receiver side, the low noise amplifier, the synthesizer, the mixer, and the ADC affect the noise floor and reduce the sensitivity. The LNA needs to provide a large linear gain and a reduced noise figure. A FoM for this device can be defined as

$$FoM_{LNA} = \frac{G_{LNA}IIP3f}{(NF-1)P_{LNA}},$$ (3.31)

where NF is the noise figure (linear scale), G_{LNA} is the gain, $IIP3$ is the third-order input intersection point, f is the operation frequency, and P_{LNA} is the power consumption. It can be observed that, considering a fixed FoM, a reduced NF and a high gain are obtained at the cost of a large power consumption. On the other side, more stringent EVM levels require higher ADC resolution, resulting also in a higher power consumption as discussed in Sect. 3.4.2. As a conclusion, dense constellations demand a higher power consumption. The trade-off between power consumption and constellation size (allowed distortion level) varies with the application. This explains why low-power applications (like IoT) use lower constellation sizes.

3.7   Energy Efficiency of Digital Compensation Techniques

As was studied in previous sections, the power amplifier and its nonlinear response are one of the main concerns in terms of system performance degradation and power consumption. There are two main alternatives to improve the energy efficiency of wireless transceivers where the power amplifier is the dominant power consumer: (a) digital predistortion (DPD) and (b) PAPR reduction methods [49, 50]. The implementation of DPD is intended to improve the PA linearity, and increase its operation range and energy efficiency. This technique allows the reduction of the input back-off, which reduces the PA power consumption. However, DPD requires an overhead that produces an increment in the consumption [13].

The use of adaptive predistorters requires a sample of the RF signal at the PA output, its downconversion to baseband and to the digital domain (an additional mixer and ADC). These additional circuits increase the power consumption of the transceiver and can be impractical for low-power systems as picocells and handsets. The energy saving obtained with the implementation of DPD needs to be evaluated including the computational cost required to implement the training and operation phases, and the power consumption of the extra component. These issues are studied in detail in Chap. 4.

The PA efficiency also depends on the PAPR of the input signal. Signals with large PAPR require the operation of the PA far from the saturation point to avoid the clipping and minimize the distortion. However, the operation in this region results in a low power efficiency because the amplifier is designed to obtain its peak efficiency when it is saturated, i.e., for low back-off.

Power efficiency values below 5% are reported for PA class A, operating with OFDM signals with 256 subcarriers. The reduction of the signal PAPR is one option to improve the PA efficiency, allowing its operation in a region close to the saturation region. As is defined in Sect. 2.5.1, the power efficiency as function of the PAPR of a PA class A can be written as $\rho = \frac{0.5}{PAPR}$. Considering that the PAPR of the signal before and after the compensation are denoted, respectively, as $PAPR_1$ and $PAPR_2$, the efficiency gain after PAPR reduction can be written as:

$$\Delta_\eta = \rho_2 - \rho_1 = 0.5 \frac{PAPR_1 - PAPR_2}{PAPR_2 PAPR_1}. \tag{3.32}$$

Considering a transmitted power P_{tx}, the consumed power is $P_{cc} = P_{tx}/\rho$, and the gain in terms of power consumption is given by

$$P_{cc_2} - P_{cc_1} = 2P_{tx}(PAPR_1 - PAPR_2). \tag{3.33}$$

From (3.33) can be observed that the energy saving is proportional to the transmitted power, what makes the implementation of PAPR reduction techniques more attractive for high-power transceivers. It is worth to mention that the implementation of PAPR reduction techniques also consumes additional processing power. By denoting the processing consumption as P_{cc_p}, the net power gain can be expressed as $P_{cc_2} - P_{cc_1} - P_{cc_p}$.

The cascade of PAPR and DPD reduction techniques is considered in new transceiver implementations [51]. The efficiency of the RF front-end including the cascade PAPR reduction-DPD can be expressed as

$$\eta_{rffe} = \frac{P_{tx}}{P_{bb} + P_{cc_p} + P_{cc_{dpd}} + P_{PA}}, \tag{3.34}$$

where P_{bb}, P_{cc_p}, $P_{cc_{dpd}}$, and P_{PA} are the power consumption of baseband, PAPR and DPD reduction, and PA blocks, respectively. The efficiency of the transceiver can be improved by minimizing the denominator of (3.34). To that purpose, the power amplifier consumption is reduced by applying PAPR and DPD. However, the implementation of these complex algorithms introduces additional consumption terms, namely P_{cc_p} and $P_{cc_{dpd}}$. The mutual influence of these parameters needs to be carefully evaluated for each application to reach an energy-efficient solution.

To illustrate the problem, the gain obtained by PAPR reduction techniques in a complete system is evaluated considering two scenarios. The Scenario 1 considers a mid-power PA with an average transmitted power of $P_{tx} = 10\,\text{W}$, whereas the Scenario 2 a low-power PA with a power of $P_{tx} = 0.2\,\text{W}$ [51]. The parameters of both scenarios are listed in Table 3.7. This example is useful to show that the implementation of baseband processing techniques is not always energy efficient. For low-power PAs, employed for low-range communication systems, the use of PAPR reduction methods increases the consumed power of the system. On the other side, for larger transmitted power, the power consumption of the transceiver is dominated by the PA and the implementation of PAPR reduction gives a large power saving.

Table 3.7 Evaluation of power saving using PAPR reduction techniques for mid- and low-power PAs

	Scenario 1 (mid-power PA)	Scenario 2 (low-power PA)
P_{tx}	10 W	0.2 W
ρ_{pa}	0.1	0.1
P_{pa}	100 W	2 W
$PAPR_1$	12 dB	12 dB
$PAPR_2$	8 dB	8 dB
ρ_{pa_A} (after PAPR reduction)	0.2	0.2
P_{pa_A} (after PAPR reduction)	50 W	1 W
P_{cc_p}	1.8 W	1.8 W
Power saving	48.2 W	−0.8 W

3.8 Summary of the Key Points

- The definitions of energy efficiency and spectral efficiency as a figure of merit of a communication systems were presented in this chapter. These parameters are key issues in the search of energy-efficient communication systems.
- Models for the power consumption of RF and baseband blocks that composed a typical transceiver were described in the chapter.
- The power consumption of long-range communication systems is a function of the power amplifier consumption. In this case, the implementation of linearization techniques is mandatory. However, when the transmitted power is lower (short-range communication), DSP and ADC converters appear as significant contributors to the power driven.
- The consumption of digital blocks needs to be considered in low-mean power devices. ADCs and digital processing block need to be carefully designed in massive MIMO applications.

References

1. G. Auer, V. Giannini, C. Desset, I. Godor, P. Skillermark, M. Olsson, M.A. Imran, D. Sabella, M.J. Gonzalez, O. Blume, A. Fehske, How much energy is needed to run a wireless network? IEEE Wireless Commun. **18**(5), 40–49 (2011)
2. Q. Wu, G.Y. Li, W. Chen, D.W.K. Ng, R. Schober, An overview of sustainable green 5G networks. IEEE Wireless Commun. **24**(4), 72–80 (2017)
3. S. Buzzi, I. Chih-Lin, T.E. Klein, H.V. Poor, C. Yang, A. Zappone, A survey of energy-efficient techniques for 5G networks and challenges ahead. IEEE J. Sel. Areas Commun. **34**(4), 697–709 (2016)
4. E. Oh, K. Son, B. Krishnamachari, Dynamic base station switching-on/off strategies for green cellular networks. IEEE Trans. Wireless Commun. **12**(5), 2126–2136 (2013)
5. P. Chang, G. Miao, Energy and spectral efficiency of cellular networks with discontinuous transmission. IEEE Trans. Wireless Commun. **16**(5), 2991–3002 (2017)

6. X. Lu, P. Wang, D. Niyato, D.I. Kim, Z. Han, Wireless networks with RF energy harvesting: a contemporary survey. IEEE Commun. Surv. Tutorials **17**(2), 757–789 (2015)
7. S. Bi, C.K. Ho, R. Zhang, Wireless powered communication: opportunities and challenges. IEEE Commun. Mag. **53**(4), 117–125 (2015)
8. M. Ariaudo, I. Fijalkow, J.-L. Gautier, M. Brandon, B. Aziz, B. Milevsky, Green radio despite "Dirty RF" front-end. EURASIP J. Wireless Commun. Netw. **2012**(1), 146 (2012)
9. J. Joung, C.K. Ho, S. Sun, Green wireless communications: a power amplifier perspective, in *Signal Information Processing Association Annual Summit and Conference (APSIPA ASC), 2012 Asia-Pacific* (2012), pp. 1–8
10. C. Desset, E. De Greef, B. Debaillie, *Power model for Today's and Future Base Stations* (2015). http://www.imec.be/powermodel
11. I. Gomez-Miguelez, V. Marojevic, A. Gelonch, Processing-to-amplifier power ratio for energy efficient communications. Electron. Lett. **48**(12), 732–734 (2012)
12. D.Y.C. Lie, J.C. Mayeda, Y. Li, J. Lopez, A review of 5G power amplifier design at cm-wave and mm-wave frequencies. Wireless Commun. Mobile Comput. **2018**, 16 (2018)
13. G.T. Watkins, K. Mimis, How not to rely on Moore's law alone: low-complexity envelope-tracking amplifiers. IEEE Microwave Mag. **19**(4), 84–94 (2018)
14. V. Camarchia, M. Pirola, R. Quaglia, S. Jee, Y. Cho, B. Kim, The Doherty power amplifier: review of recent solutions and trends. IEEE Trans. Microwave Theory Tech. **63**(2), 559–571 (2015)
15. J. Kim, J. Son, S. Jee, S. Kim, B. Kim, Optimization of envelope tracking power amplifier for base-station applications. IEEE Trans. Microwave Theory Tech. **61**(4), 1620–1627 (2013)
16. J. Jeong, D.F. Kimball, M. Kwak, C. Hsia, P. Draxler, P.M. Asbeck, Wideband envelope tracking power amplifier with reduced bandwidth power supply waveform, in *2009 IEEE MTT-S International Microwave Symposium Digest* (2009), pp. 1381–1384
17. D. Kang, B. Park, D. Kim, J. Kim, Y. Cho, B. Kim, Envelope-tracking CMOS power amplifier module for LTE applications. IEEE Trans. Microwave Theory Tech. **61**(10), 3763–3773 (2013)
18. G.T. Watkins, K. Mimis, A 65% efficient envelope tracking radio-frequency power amplifier for orthogonal frequency division multiplex. IET Microwaves Antennas Propag. **9**(7), 676–681 (2015)
19. F. Balteanu, H. Modi, S. Khesbak, S. Drogi, P. DiCarlo, Envelope tracking LTE multimode power amplifier with 44% overall efficiency, in *2017 IEEE Asia Pacific Microwave Conference (APMC)* (2017), pp. 37–40
20. Q. Jin, X. Ruan, X. Ren, Y. Wang, Y. Leng, C.K. Tse, Series/parallel form switch-linear hybrid envelope-tracking power supply to achieve high efficiency. IEEE Trans. Ind. Electron. **64**(1), 244–252 (2017)
21. U. Karthaus, D. Sukumaran, S. Tontisirin, S. Ahles, A. Elmaghraby, L. Schmidt, H. Wagner, Fully integrated 39 dBm, 3-stage Doherty PA MMIC in a low-voltage GaAs HBT technology. IEEE Microwave Wireless Compon. Lett. **22**(2), 94–96 (2012)
22. A.M. Mahmoud Mohamed, S. Boumaiza, R.R. Mansour, Doherty power amplifier with enhanced efficiency at extended operating average power levels. IEEE Trans. Microwave Theory Tech. **61**(12), 4179–4187 (2013)
23. K. Nakatani, S. Shinjo, S. Miwa, R. Ma, K. Yamanaka, 3.0–3.6 GHz wideband, over 46% average efficiency GaN Doherty power amplifier with frequency dependency compensating circuit, in *2017 IEEE Radio Wireless Week (RWW)* (2017)
24. I.V. Singh, M.S. Alam, Cascode mixer for multiband wireless, in *IMPACT-2013* (2013), pp. 180–184
25. Energy efficiency analysis of the reference systems, areas of improvements and target breakdown, in *EARTH project deliverable, D2.3* (2012)
26. Distributed and centralized baseband processing algorithms, architectures, and platforms, in *MAMMOET project deliverable, D3.2* (2015)
27. C. Desset, B. Debaillie, F. Louagie, Modeling the hardware power consumption of large scale antenna systems, in *2014 IEEE Online Conference on Green Communications (OnlineGreen-Comm)* (2014), pp. 1–6

28. M. Lauridsen, L. Noël, T.B. Sorensen, P. Mogensen, An empirical LTE smartphone power model with a view to energy efficiency evolution. Intel Technol. J. **18**, 172–193 (2014)
29. M. Lauridsen, P. Mogensen, L. Noel, Empirical LTE smartphone power model with DRX operation for system level simulations, in *2013 IEEE 78th Vehicular Technology Conference (VTC Fall)* (2013), pp. 1–6
30. A. Mammela, A. Anttonen, Why will computing power need particular attention in future wireless devices? IEEE Circuits Syst. Mag. **17**(1), 12–26 (2017)
31. L.G. Baltar, F. Schaich, M. Renfors, J.A. Nossek, Computational complexity analysis of advanced physical layers based on multicarrier modulation, in *2011 Future Network Mobile Summit* (2011), pp. 1–8
32. C. Desset, A. Fort, Selection of channel coding for low-power wireless systems, in *The 57th IEEE Semiannual Vehicular Technology Conference, 2003. VTC 2003-Spring*, vol 3 (2003), pp. 1920–1924
33. A. Bhise, P. Vyavahare, Complexity analysis of iterative decoders in mobile communication systems. Int. J. Inf. Electron. Eng. **4**(2), 121–128 (2014)
34. B. Murmann, The race for the extra decibel: a brief review of current ADC performance trajectories. IEEE Solid-State Circuits Mag. **7**(3), 58–66 (2015)
35. S. Krone, G. Fettweis, Energy-efficient A/D conversion in wideband communications receivers, in *2011 IEEE Vehicular Technology Conference (VTC Fall)* (2011), pp. 1–5
36. T. Sundstrom, B. Murmann, C. Svensson, Power dissipation bounds for high-speed Nyquist analog-to-digital converters. IEEE Trans. Circuits Syst. I Regul. Pap. **56**(3), 509–518 (2009)
37. Y. Chen, J.A. Nossek, A. Mezghani, Circuit-aware cognitive radios for energy-efficient communications. IEEE Wireless Commun. Lett. **2**(3), 323–326 (2013)
38. A. Mezghani, J.A. Nossek, Power efficiency in communication systems from a circuit perspective, in *2011 IEEE International Symposium of Circuits and Systems (ISCAS)* (2011), pp. 1896–1899
39. A. Mezghani, J.A. Nossek, Modeling and minimization of transceiver power consumption in wireless networks, in *2011 International ITG Workshop on Smart Antennas* (2011), pp. 1–8
40. E. Björnson, L. Sanguinetti, J. Hoydis, M. Debbah, Optimal design of energy-efficient multi-user MIMO systems: is massive MIMO the answer? IEEE Trans. Wireless Commun. **14**(6), 3059–3075 (2015)
41. S. Krone, G. Fettweis, Energy-efficient A/D conversion in wideband communications receivers, in *2011 IEEE Vehicular Technology Conference (VTC Fall)* (2011), pp. 1–5
42. M. Sarajli, L. Liu, O. Edfors, When are low resolution ADCs energy efficient in massive MIMO? IEEE Access **5**, 14837–14853 (2017)
43. S. Moon, I. Kim, D. Kam, D. Jee, J. Choi, Y. Lee, Massive MIMO systems with low-resolution ADCs: baseband energy consumption vs. symbol detection performance. IEEE Access **7**, 6650–6660 (2019)
44. E. Bjornson, M. Matthaiou, M. Debbah, Circuit-aware design of energy-efficient massive MIMO systems, in *2014 6th International Symposium on Communications, Control and Signal Processing (ISCCSP)* (2014), pp. 101–104
45. E. Björnson, M. Matthaiou, M. Debbah, Massive MIMO systems with hardware-constrained base stations, in *2014 IEEE International Conference on Acoustics, Speech and Signal Processing (ICASSP)* (2014), pp. 3142–3146
46. F. Sohrabi, W. Yu, Hybrid digital and analog beamforming design for large-scale antenna arrays. IEEE J. Sel. Top. Signal Process. **10**(3), 501–513 (2016)
47. I. Ahmed, H. Khammari, A. Shahid, A. Musa, K.S. Kim, E. De Poorter, I. Moerman, A survey on hybrid beamforming techniques in 5G: architecture and system model perspectives. IEEE Commun. Surv. Tutorials **20**(4), 3060–3097 (2018)
48. F. Gregorio, J. Cousseau, S. Werner, T. Riihonen, R. Wichman, EVM analysis for broadband OFDM direct-conversion transmitters. IEEE Trans. Veh. Technol. **62**(7), 3443–3451 (2013)
49. S.H. Han, J.H. Lee, An overview of peak-to-average power ratio reduction techniques for multicarrier transmission. IEEE Trans. Wireless Commun. **12**(2), 56–65 (2005)

50. A.M. Rateb, M. Labana, An optimal low complexity PAPR reduction technique for next generation OFDM systems. IEEE Access **7**, 16406–16420 (2019)
51. A.N. Lozhkin, T. Maniwa, M. Shimizu, Rf front-end architecture for 5G, in *2018 IEEE 29th Annual International Symposium on Personal, Indoor and Mobile Radio Communications (PIMRC)* (2018), pp. 1–6

Part II
Digital Compensation Techniques

This second part of the book begins with Chap. 4, where power amplifier linearization techniques are worked. In-band distortion and out-of-band distortion are parametrized with figures of merit to measure degrees of nonlinearity. Basic (transmitter side) pre-distortion techniques aiming linearization are also introduced. Receiver side compensation techniques are also considered. Finally, a case of study (class AB and envelope–tracking PA) is introduced in this chapter. Chapter 5 considers different classes of ADC when considering the tradeoff between conversion speed and consumption: flash, successive approximations, sigma–delta, and combined structures. Considering the possible modeled impairments (integral nonlinear and dynamic nonlinear models) different compensation techniques are considered: model inversion, mismatch error compensation, etc. This part concludes with Chap. 6 where frequency offset and phase noise effects are described and discussed. Specific cases where these impairments are critical are discussed in detail. Also, estimation and compensation of carrier frequency offset in the DL and UL are considered.

Part II
Digital Compensation Techniques

Chapter 4
Power Amplifiers

Abstract Conventionally, the power amplifier is considered the most power demanding device in a wireless transceiver. Following this assumption, improvements in the power efficiency of the power amplifier (PA) are directly reflected in the power consumption of the overall system. Considering the power consumption of a LTE macrocell base station, the PA drains around 57% of the total power, while the baseband processing requires only 13%. However, in a femtocell, the PA consumption represents only 22% of the total power and the portion of the digital block demands 47%. For the case of power limited devices, as a LTE mobile phone, the PA dominates the overall power consumption requiring 44% of the total available power. We can infer from these values that for high/medium power systems, as macrocell base stations, the implementation of linearization techniques or peak-to-average power ratio (PAPR) reduction methods are mandatory. In that case, their implementation allows to relax the linearity constraints and improve the power efficiency. A substantial energy saving can thus be obtained in that scenario. On the other hand, for low power transceivers, the trade-off between the energy saved by optimizing the PA operation point and the energy required to implement predistortion techniques needs to be carefully evaluated. In this chapter, we address the problem of power consumption in power amplifiers and their effects over energy efficiency and performance of a wireless transceiver. The trade-off between the allowed power amplifier distortion and the system power efficiency is studied. Several nonlinear distortion compensation techniques that can be applied either in the receiver or the transmitter side are introduced and their performance is studied for several scenarios.

4.1 Power Amplifiers and Multicarrier Signals

The large PAPR of multicarrier signal imposes the operation of the PA in a low-power efficiency region. The traditional approach to amplify OFDM signals is to back-off class A or class AB amplifiers until the distortion level is within acceptable margins. In that way, however, poor power efficiency solutions are obtained [1].

© Springer Nature Switzerland AG 2020

F. Gregorio et al., *Signal Processing Techniques for Power Efficient Wireless Communication Systems*, Signals and Communication Technology, https://doi.org/10.1007/978-3-030-32437-7_4

While the input stages of a base station PA are commonly biased in class A mode, the output stage operates in a less linear but more efficient class AB or B mode [2]. The efficiency of a class AB stage actually ranges from 5 to 10% when operated under back-off conditions in OFDM context [3–5].

To maintain satisfactory levels of distortion while increasing the power efficiency, external compensation techniques must be employed. A number of techniques exists to linearize the operating region of the PA, see, e.g., [6]. Examples of linearization techniques are as follows: feed-forward linearizer, feedback linearizer, envelope elimination and restoration, and digital predistortion (DPD) [7, 8]. The latter, a very cost effective solution, is the main focus of this chapter. PAPR reduction techniques are also an alternative to avoid the operation of the PA in the nonlinear region reducing the distortion [9].

Constant search for power efficient solutions motivates the revival of amplifier configurations as envelope tracking and Doherty. Envelope tracking (ET) amplifiers improve the power efficiency by varying the amplifier supply voltage synchronized with the envelope of the OFDM signal. Average power efficiency levels around 45% are reported for ET amplifiers for mobile LTE applications. ET PA can achieve an average efficiency of over 60% for LTE base stations (tested with PAPR around 6.5 dB). However, ET has an inherent nonlinearity associated with the gain variation as the supply voltage changes. The power amplifier output features such as power, efficiency, gain, and phase now depend on two control inputs: RF input power and supply voltage. This motivate the development of novel predistortion techniques.

Memory polynomial (MP) predistorter [10] and the generalized memory polynomial (GMP) [11] are efficient and robust models employed for broadband PA linearization [12], where memory effects are significant and must be considered. However, large polynomial order is needed to linearize the response of certain PAs such as class AB, increasing implementation complexity. Also, the use of large polynomial order to improve modeling representation leads to ill-conditioned DPD polynomial coefficients, reducing the quality of the estimates and the generalization properties of the model. Furthermore, the increasing ratio between the magnitude of the maximum and minimum values of the polynomial coefficients as the polynomial order grows [13], also leads to degradation of the linearization performance due to limited-resolution DSP hardware implementation.

4.1.1 Operation Point: Power Consumption vs Distortion Trade-Off

To avoid operation of the power amplifier in the nonlinear region, the input signal is scaled down (a power back-off is applied) to fit into the linear part of the PA transfer characteristic. The application of the back-off results in a significant increase in the PA consumption compared to the power of the amplified signal, i.e., the PA design is over-dimensioned, resulting in a reduced power efficiency [14]. For this reason, back-off is restricted to applications with low requirements of power efficiency.

There are two quantities to measure the level of back-off: (a) Input back-off (IBO) and (b) Output back-off (OBO). They are defined as:

$$\text{IBO} = 10 \log \frac{P_{i,sat}}{P_i} \tag{4.1}$$

and

$$\text{OBO} = 10 \log \frac{P_{o,sat}}{P_o}, \tag{4.2}$$

where $P_{i,sat}$ and $P_{o,sat}$ are the input and output saturation powers and P_i and P_o are the mean power of the input and output PA signals, respectively. Either OBO or IBO can be used to specify the PA operation point, e.g., OBO is useful to define the power being generated out of the amplifier compared with the maximal available power. Figure 4.1 illustrates the definition of OBO and IBO parameters for a generic PA.

As mentioned above, the use of back-off to linearize the system comes at the cost of a decreased power efficiency. Operating the amplifier in the saturation region allows for a maximum power efficiency while sacrificing linearity. This trade-off is more severe when high-power amplifiers are considered [14].

Figure 4.2 illustrates the performance of a nonlinear amplifier operating with input back-off of 2 and 5 dB, respectively. The OFDM transmitter has $N = 512$ subcarriers and 16-QAM modulation. The PA is modeled as a memoryless solid state power amplifier (SSPA), as defined in Chap. 2, with smoothness factor $p = 2$.

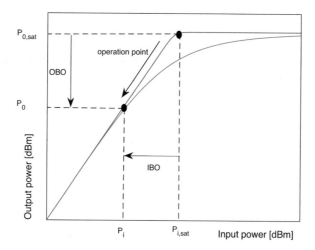

Fig. 4.1 Output and input back-off definition

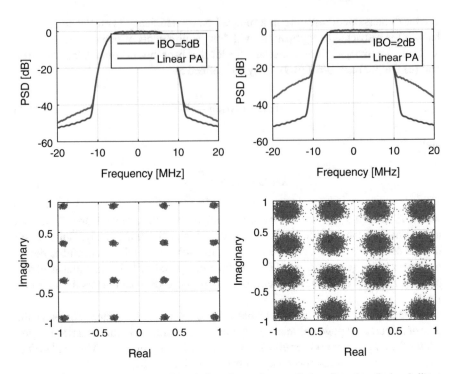

Fig. 4.2 Spectral regrowth (out-of-band distortion and constellation distortion (in-band distortion)) in function of power amplifier operation point

It can be observed that when the PA is operating with low back-off (IBO = 2 dB), the level of in-band distortion and spectral regrowth raise to large values making the system not suitable for practical implementations. Spectrum broadening and constellation compression plus distortion are observed in the PSD and constellation figures, respectively.

If the operation is moved away from the saturation region (IBO = 5 dB), the PA will operate in the linear region minimizing the in-band and out-of-band distortions but results in a solution with low power efficiency.

The linearity and power efficiency constraints motivate the development of more advanced compensation techniques as predistorters (transmitter side) and iterative interference remotion (receiver side), as discussed in the next sections.

4.2   Linearization Techniques

The compensation of nonlinear distortion generated by the transmitter PA can be implemented either at the transmitter or the receiver side [6].

Reduction of PA nonlinear distortion is often preferable in the transmitter to avoid the problems due to the wireless channel, since it introduces additional memory and has time varying characteristics [15]. On the other hand, transmitter compensation techniques, like predistorters and PAPR reduction, may render on implementation complexity that is prohibitive in small handsets (cellular telephony) or subscriber units (Wi-Fi and WiMAX).

In the transmitter side methods, the signal to be transmitted is modified before the PA. The well known methods for that purpose are predistortion and PAPR reduction techniques. A digital predistorter (DPD) aims to model the inverse of the PA non-linear response and is placed before the PA such that the cascade PD-PA produces a linearly amplified signal [16]. In case of narrowband signals and low/medium power amplifiers, a memoryless DPD structure, where the current output depends only on the current input, is the best option in terms of implementation complexity and performance. In that case, the DPD is described by a static nonlinear function often implemented with polynomial models [17]. In broadband OFDM implementations, however, memory effects that appear in the PA response need to be considered in the DPD design. Volterra, Wiener, Wiener–Hammerstein, and memory polynomial models are generally employed for these cases of DPD design [18, 19].

Receiver-side compensation can be justified for uplink transmission moving the processing task to the base station where higher computationally complexity is allowed. As a consequence, mobile terminals are kept simple and power-efficient [20, 21]. It is worth to mention that receiver-side compensation techniques must deal with the estimation problems associated with the channel (i.e., memory effects and time-varying characteristics).

Receiver-side compensation techniques could be a good alternative for scenarios such as IoT-based uplink where an enhanced coverage power-efficient terminal transmits to a high-quality receiver [22].

Nonlinear PA distortion compensation at the receiver enables a reduced transmitter complexity when compared with a transmitter compensation solution. Instead, the demanding processing is moved to the receiver. The choice between a receiver or a transmitter implementation depends on system complexity that can be supported.

It is important to remark that receiver compensation techniques only compensate the effects of in-band distortion that causes bit error rate (BER) degradation. Therefore, the PA needs to be biased to operate in a region where the spectral regrowth verifies the spectral mask imposed by the standards. Another alternative is the combination of receiver compensation techniques and transmitter compensation techniques that can be useful to obtain suitable BER levels and a reduced spectral regrowth, at a moderate computational cost for the transmitter.

Receiver compensation techniques include postdistortion methods [23–27], and nonlinear equalization [28–30] both developed for single-carriers systems, and iterative detection methods [31–37] that can be considered for OFDM systems. Power amplifier nonlinearity cancellation (PANC) introduced in [20] can be applied in several scenarios, including relay systems [38]. A version of PANC adapted for FBMC/OQAM system is presented in [39]. It is also addressed in [40] considering single carrier frequency division multiple access systems (SCFDMA). A variation of PANC, called reconstruction of distorted signals (RODS), is presented in [41] reaching a very good performance for PA with severe memory effects. The identification of PA and channel response, required for iterative detection techniques as PANC and RODS, are studied in several research articles as [22, 42].

4.3 Figures of Merit: In-Band and Out-of-Band Distortion

The nonlinear response of the RF front-end creates in-band and out-of-band distortion. Several figures of merit can be considered to evaluate the performance of the linearization techniques. Specifically: Error vector magnitude (EVM), defined in Eq. (2.12), that quantifies in-band distortion and is directly related to the bit error rate (BER), and adjacent channel power ratio (ACPR) that measures the effects of the out-of-band distortion on adjacent channels. EVM and ACPR are employed for TX side linearization techniques.

Total degradation is the figure of merit to evaluate when receiver-side nonlinearities remotion techniques are considered.

- Adjacent channel power ratio (ACPR)
  The out-of-band distortion is directly related to the power amplifier operation point. The out-of-band emission increases when the PA is driven into its nonlinear operation region, which allows larger power efficiency. The adjacent channel power ratio is employed to characterize the spectral regrowth and is defined as:

$$ACPR = 10 \log \left(\frac{\int_{f_{ad}} Y(f) df}{\int_{f_{main}} Y(f) df} \right), \qquad (4.3)$$

  where $Y(f)$ is the power spectral density (PSD) at the output of the linearized PA. f_{ad} and f_{main} define the frequency bands of the adjacent channel and the main channel, respectively.
- Total degradation (TD)
  Total degradation is a measure for the balance between the level of degradation and power efficiency, which is defined as:

$$TD_{dB} = SNR_{NL}[dB] - SNR_L[dB] + IBO[dB], \qquad (4.4)$$

where $= SNR_{NL}[dB]$ is the SNR required to obtain a target BER in the presence of PA nonlinearities with a fixed IBO, and SNR_L expresses the SNR required in the case of a linear PA.

4.4 Digital Predistortion Techniques

The predistorter (PD) is a nonlinear device placed before the nonlinear PA such that a linearly amplified signal is obtained at the output of the system (see Fig. 4.3). If the PD models the inverse of the PA nonlinear response, the nonlinear components are cancelled obtaining a scaled version (linear gain) of the input signal.

Three different categories of PD can be distinguished [2]: (1) RF predistorter which works directly at the carrier frequency, (2) Intermediate frequency (IF) predistorter that operates at the IF band, and (3) baseband predistorter where the predistortion is moved to the baseband frequency allowing the application of DSP techniques. Nowadays, digital baseband predistorters are the preferred option.

Memoryless digital predistorter (DPD), in which the current output depends only on the current input, are the most simple predistorters structures. The PA response is compensated using an expanding nonlinearity corresponding to the PA inverse transfer function. The memoryless predistortion techniques that have been developed during the last decade are either look-up-table (LUT) based or parameter based.

In the LUT approach a complete input–output mapping is stored in a memory device and the original input constellation points are mapped to the desired location adjusting the input signal by the inverse characteristic of the amplifier [43–46]. The drawback of the LUT approach is the memory requirements, especially when the LUT is extended to include memory effects. In the parametric approach, the predistorter may be described by a nonlinear function, polynomial model, piecewise linear model, or Volterra model [47, 48]. Parametric implementations for memoryless PA are proposed in e.g., [49, 50].

To estimate the PD parameters, one of the following methods are employed:

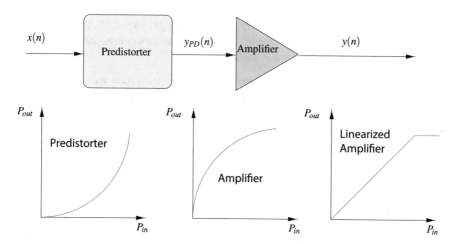

Fig. 4.3 Block diagram describing the principle of predistorter. The PD modifies the input signal $x(n)$ such that the resultant output $y(n)$ is a linear amplified version of the input signal

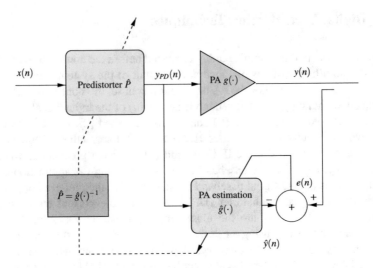

Fig. 4.4 Predistorter implementation using direct learning structure

- Direct learning: The PA model is estimated first and used later to find the inverse [51]. The block diagram of the direct adaptation structure is illustrated in Fig. 4.4, where the PA input $y_{PD}(n)$ and PA output $y(n)$ are used to calculate an estimate of the PA model $\hat{g}(\cdot)$. This estimate is then used to calculate the PA inverse P.
- Indirect learning adaptation: In order to avoid the calculation of the inverse of the power amplifier, the indirect learning method is adopted [17]. This method is a popular approach to estimate the predistorter model. In this structure the predistorter is directly designed eliminating the PA estimation step. In the structure depicted in Fig. 4.5, the predistorter parameters in the training branch A are obtained by minimizing the error signal $e(n)$ expressed by the training branch output signal $\hat{y}_{PD}(n)$ and the predistorter output signal $y_{PD}(n)$. The parameters obtained in the training branch are directly copied to the predistorter, avoiding the inverse calculation process.

Memory effects appear in broadband applications such as multicarrier systems (electrical memory effects) or for high-power amplifiers (thermal memory effects), and must be considered in the predistorter design [11]. Volterra series are employed for nonlinear memory modeling. In the case of direct learning, implementations based in Volterra series are reported in [52–54]. A truncated Volterra series model based on the P-th inverse and fixed-point approaches have been proposed for PD structures achieving good performance [55, 56]. However, these methods have high implementation complexity. Preliminary work of indirect-learning Volterra predistorter was done in [17].

The main drawback of Volterra based PD is the high implementation complexity related with the high number of parameters. Slow convergence is also another fundamental issue that constraints its implementation in real systems.

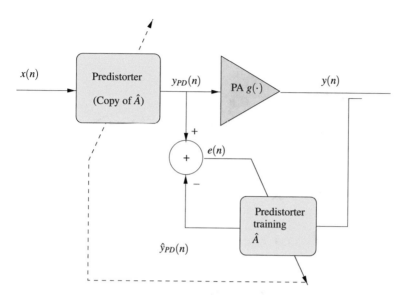

Fig. 4.5 Predistorter implementation using indirect learning structure

To reduce the number of parameters related with Volterra models while still keeping a good modeling capability, Wiener, Hammerstein, and memory polynomial models are good alternatives. The Wiener system has been developed for PA predistortion in several papers [15, 51, 57, 58]. Hammerstein and memory polynomial predistorters are evaluated in [59]. In [18] a memory polynomial model is proposed for use in an indirect learning structure. The proposed PD maintains a good performance when is evaluated for different PA models. The numerical problems associated with the estimation of predistorter coefficients in conventional polynomial models can be alleviated by using orthogonal polynomials models [60]. A generalized memory polynomial model that outperform conventional memory polynomial is presented in [11] and employed with good results in [61–63].

4.4.1   Baseband Predistortion Techniques: Implementation Issues

The implementation of a digital predistorter requires the addition of an observation path employed for the estimation of the PD parameters. The sensed RF signal at the output of the PA is downconverted to baseband and sampled. The digital samples are processed and predistortion parameters are obtained. These parameters modify the baseband I/Q signal in such a way that the distortion created by the PA is minimized. The observation path adds some additional components as RF sampler, attenuator, downconverter, and ADC. These components also introduce some distortion that

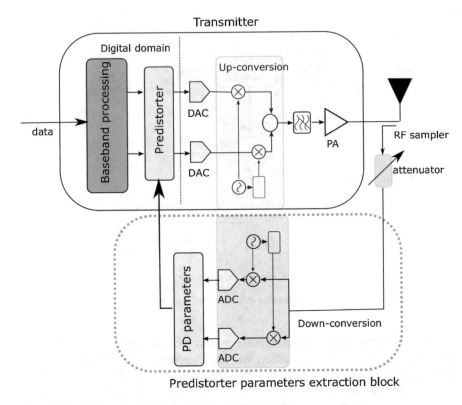

Fig. 4.6 Block diagram of digital predistorter implementation

needs to be taken into account when the DPD coefficients are calculated. Figure 4.6 illustrates a block diagram of a transmitter with a digital predistorter including the PD coefficient extraction block and the main path.

The main issues associated with the implementation of baseband predistorters are the required oversampling, the added complexity, and the power consumption. The added power consumption is determined by the ADC sampling rate and resolution. Considering a nonlinear PA with a fifth-order dominant nonlinearity, a conventional DPD architecture must observe the PA output signal at 5 times of the input bandwidth in order to capture the nonlinear effects. For example, in case of a LTE-advanced signal that utilizes a transmission bandwidth up to 100 MHz, a sampling rate of 500 MHz must be employed. Moreover, a large resolution is required to improve the linearization capability. High sampling rate and large resolution require high cost and high power consumption ADCs that are not affordable in real implementations [16]. Several solutions are available to solve the challenging DPD design. They allow to obtain good performance under band-limited situations operating with a reduced sampling rate [64–66].

4.5 Receiver-Side Compensation Techniques

Receiver-side compensation techniques can be applied in two different scenarios: (a) to remove the nonlinear distortion generated by deliberate clipping implemented to increase the PA dynamic range and its efficiency and (b) to remove the nonlinear distortion created by a nonlinear transfer function without clipping. In the first case, only the clipping level knowledge is required for the implementation of compensation method [31, 34]. In the second scenario, the knowledge of the PA transfer function is required to remove the distortion. This more challenging scenario is addressed in [20, 22, 42]. In the following, an example of the two approaches is presented.

4.5.1 Decision-Aided Reconstruction of Clipped Signals

Decision-aided reconstruction (DAR) technique proposed in [31] addressed the reconstruction of clipped OFDM signals. Considering a soft limiter PA, the clipped signal $x_g(n)$ can be written as:

$$x_g(n) = \begin{cases} x(n), & |x(n)| < A_s \\ A_s, & |x(n)| > A_s \end{cases}, \qquad (4.5)$$

where $x(n)$ is the time domain input sequence, and A_s is the output saturation voltage (clipping level) of the PA. The clipped signal is transmitted through the channel. The received signal in frequency domain (after DFT and cyclic prefix remotion) is given by

$$Y(n, k) = H(n, k) K_L X(n, k) + U(n, k), \qquad (4.6)$$

where $H(n, k)$ is the complex channel gain at subcarrier k, $U(n, k) = H(n, k) W_D(n, k) + W(n, k)$ is an additive distortion term composed by the clipping noise and channel AWGN, and K_L is the gain-compression factor. For large clipping levels, $K_L \cong 1$ and is neglected in the reconstruction algorithm as summarized below:

1. The soft estimate of the clipped transmitted signal is obtained as $\hat{X}_g(n, k) = \frac{Y(n,k)}{H(n,k)}$ for $k = 0, \ldots, N - 1$.
2. For iteration m, the symbol decisions $\hat{X}^{(m)}(n, k)$ are found according to

$$\hat{X}^{(m)}(n, k) = \arg\min_X \left| Y^{(m)}(n, k) - H(n, k) X(n, k) \right|, \quad 0 \le k \le N - 1,$$
$$(4.7)$$

where $\hat{Y}^{(0)}(n, k) = Y(n, k)$.

3. The soft estimates $\{\hat{X}_g(n,k)\}_{k=0}^{N-1}$ in Step 1 and symbol decisions $\{\hat{X}^{(m)}(n,k)\}_{k=0}^{N-1}$ are converted to their corresponding time domain sequences $\{\hat{x}_g(n,i)\}_{i=0}^{N-1}$ and $\{\hat{x}^{(m)}(n,i)\}_{i=0}^{N-1}$ using the IDFT. Clipped samples are detected and a new sequence is generated $\{\tilde{x}_g^{(m)}(n,i)\}_{i=0}^{N-1}$ according to

$$
\tilde{x}_g^{(m)}(n,i) = \begin{cases} \hat{x}_g(n,i), & |\hat{x}^{(m)}(n,i)| \leq A_s \\ \hat{x}^{(m)}(n,i), & |\hat{x}^{(m)}(n,i)| > A_s \end{cases} \tag{4.8}
$$

for $0 \leq i \leq N-1$.
4. The sequence $\{\tilde{x}_g^{(m)}(n,i)\}_{i=0}^{N-1}$ is converted back to frequency domain obtaining $\{\tilde{X}_g^{(m)}(n,k)\}_{i=0}^{N-1}$.
5. The index number is increased $m = m+1$ and $Y^{(m)}(n,k)$ is determined by

$$
Y^{(m)}(n,k) = H(n,k)\tilde{X}_g^{(m-1)}(n,k) \tag{4.9}
$$

6. Go to Step 2 to refine the symbol estimates.

The DAR technique [31] reaches good performance for clipping levels larger than 4 dB considering a static channel. When the clipping level is reduced, the decision errors degrade the performance. This is a problem associated with most decision-based techniques. DAR techniques are useful to recover signal affected by clipping operation. However, DAR is not able to cope with the nonlinear response of the PA in the operation region, i.e., when the input power is small compared to that of saturation.

4.5.2 Power Amplifier Nonlinearity Cancellation (PANC)

This section presents power amplifier nonlinearity cancellation (PANC) technique that is an iterative approach for removing the nonlinear PA effects at the receiver side. The proposed technique deals with the challenging case of nonlinear PAs with memory and time-varying transmission channels.

The nonlinear PA with memory is modeled using a Wiener–Hammerstein structure [11]. The model is frequently used to model broadband PAs and is formed by a cascade of a linear filter $A(z)$, a nonlinear static function $g[\cdot]$, and another linear filter $B(z)$. The filters, $A(z)$ and $B(z)$, are here modeled as FIR filters of orders N_a and N_b, respectively, and the static nonlinearity is modeled with a polynomial.

The transmitted signal $z(n)$ can now be expressed as:

$$
z(n) = \sum_{m=0}^{N_b-1} b_m g[s(n-m)], \tag{4.10}
$$

where

$$s(n) = \sum_{m=0}^{N_a-1} a_m x(n-m). \tag{4.11}$$

The output of the nonlinear static block $g[\cdot]$ is modeled using the additive model [67] given by

$$g[s(n)] = K_L s(n) + w_d(n)$$
$$E[s^*(n)w_d(n)] = 0. \tag{4.12}$$

The first term in (4.12) is just a scaled version of the input signal ($K_L \leq 1$), while $w_d(n)$ is an additive distortion term.

Replacing (4.12) in (4.10) and assuming the effective channel length $N_t = N_a + N_b + N_h$ is shorter than the cyclic prefix length, the received signal at subcarrier k can be written as:

$$Y(k) = C(k)B(k)\left[K_L A(k)X(k) + W_d(k)\right] + W(k), \quad k = 0, \ldots, N-1 \tag{4.13}$$

where for subcarrier k, $W(k)$ is the additive noise, $W_d(k)$ is the nonlinear distortion (DFT of $\{w_d(n)\}_{n=0}^{N-1}$), and $A(k)$, $B(k)$, and $C(k)$ denote the responses of the linear filters of the PA model and the wireless channel, respectively.

To detect $X(k)$ from the received signal in (4.13), the following decision rule can be considered

$$\hat{X}(k) = \arg\min_{X(k)} \left\{ \left\| H_L(k)\left[X(k) - \frac{Y(k)}{H_L(k)} + \frac{W_d(k)}{A(k)K_L}\right] \right\|^2 \right\} \quad k = 0, 1, \ldots, N-1 \tag{4.14}$$

where

$$H_L(k) = K_L A(k)B(k)C(k). \tag{4.15}$$

It is assumed that $H_L(k)$, $A(k)$, and $W_d(k)$ are all known at the receiver, the estimate is given by the symbol $X(k)$ with the minimum distance to $Y(k)/H_L(k) - W_d(k)/(A(k)K_L)$. The distortion $W_d(k)$ can be estimated (assuming the knowledge of the static nonlinear function $g[\cdot]$) as:

$$w_d(n) = g[s(n)] - K_L s(n), \quad n = 0, \ldots, N-1$$
$$W_d(k) = \frac{1}{N} \sum_{n=0}^{N-1} w_d(n)e^{-j2\pi n \frac{k}{N}}, \quad k = 0, 1, \ldots, N-1. \tag{4.16}$$

From (4.11) can be observed that $s(n)$ depends on both $\{X(k)\}_{k=0}^{N-1}$ and $A(z)$. Following the approach in [32], an iterative technique that provides tentative decisions $\hat{X}(k)$, the distortion term can be estimated and removed. PANC technique is illustrated in Fig. 4.7, and the algorithm is summarized in Table 4.1.

The PANC technique requires, in addition to the tentative decisions $\hat{X}(k)$, the estimates $\hat{H}_L(k)$, $\hat{A}(k)$, and $\hat{g}[\cdot]$. A receiver-based estimation method of the PA model parameters and wireless channel is presented in [42].

The performance of the PANC technique is illustrated using an OFDM system with 16-QAM modulation on $N = 512$ subcarriers and a cyclic prefix of 64. The system operates with a carrier frequency is $f_c = 5\,\text{GHz}$ and bandwidth of $B = 20\,\text{MHz}$. The channel is Rayleigh fading with independent propagation paths, each generated according to a Jake's Doppler spectrum. The power loss and delay profiles

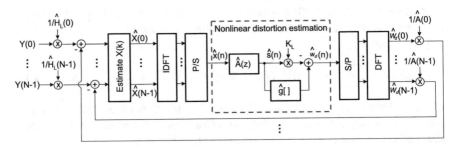

Fig. 4.7 Power amplifier nonlinearity cancellation (PANC) with the proposed PA distortion model

Table 4.1 Power amplifier nonlinearity cancellation (PANC)

PANC
Initialization:
$\hat{W}_d^{(0)}(k) = 0, \quad k = 0, \ldots, N-1$
for $m = 1$ to $I_{\max}$
{
$\quad$ Symbol decoding:
$\quad \hat{X}^{(m)}(k) = \arg\min_{X(k)} \left\{ \left\| H_L(k) \left[X(k) - \frac{Y(k)}{H_L(k)} + \frac{\hat{W}_d^{(m-1)}(k)}{A(k)} \right] \right\|^2 \right\}$
$\quad$ Time domain:
$\quad \hat{x}^{(m)}(n) = \frac{1}{\sqrt{N}} \sum_{k=0}^{N-1} X^{(m)}(k) e^{j2\pi k \frac{n}{N}}$
$\quad$ Estimate $s(n)$:
$\quad \hat{s}^{(m)}(n) = \sum_{l=0}^{L_a} a_l \hat{x}^{(m)}(n-l)$
$\quad$ Estimate distortion $w_d(n)$:
$\quad \hat{w}_d^{(m)}(n) = g[\hat{s}^{(m)}(n)] - K_L \hat{s}^{(m)}(n)$
$\quad$ Distortion in frequency domain $W_d(k)$
$\quad \hat{W}_d^{(m)}(k) = \frac{1}{\sqrt{N}} \sum_{n=0}^{N-1} w_d^{(m)}(n) e^{-j2\pi n \frac{k}{N}}$
}

of the channel are $[0, -1, -3, -9]$ dB and $[0, 1, 2, 3]$ µs corresponding to an urban scenario and a mobile speed of 2 km/h. The evaluation includes the receiver side PA model estimation proposed in [42]. The unknown wideband PA is modeled as a Wiener–Hammerstein system, where the linear filters (IIR) are given by [60]

$$A(z) = \frac{1 + 0.1z^{-2}}{1 - 0.1z^{-1}} \text{ and } B(z) = \frac{1 - 0.1z^{-1}}{1 - 0.2z^{-1}} \qquad (4.17)$$

and the static nonlinearity model as a solid state power amplifier (SSPA) employing a Saleh model [68], i.e.,

$$g[x(n)] = \frac{|x(n)|}{\left[1 + \left(\frac{|x(n)|}{A_s}\right)^{2p}\right]^{1/p}} \exp[j\angle x(n)], \qquad (4.18)$$

where the parameter p adjusts the smoothness of the transition from the linear region to the saturation region, and A_s is the amplifier input saturation point. The results are evaluated for different clipping levels γ defined as:

$$\gamma = \frac{A_s}{\sqrt{E_n\{|x(n)|^2\}}}, \qquad (4.19)$$

where $\sqrt{E_n\{|x(n)|^2\}}$ is the RMS value of the OFDM signal. The simulation study suggests that two or three iterations (I_{max} in Table 4.1) are usually sufficient.

Figure 4.8 shows the BER considering a clipping level of $\gamma = 3$ dB. Curves for a linear PA with channel estimation and channel state information (CSI), nonlinear PA, and PANC with perfect PA model estimation are included for reference. It can be observed that the PANC technique employing receiver-based model estimation reaches almost identically result to a solution having perfect knowledge of the PA model. The curves are obtained considering 3 PANC iterations.

Total degradation curves for a nonlinear PA with and without PANC shown in Fig. 4.9. We see that the best operating point for a system with PANC is IBO = 1 dB with a TD = 1 dB. Without applying PANC, the optimal point is IBO = 3.5 dB with a TD = 3.5 dB. Assuming a class A PA, the optimal points define a power efficiency of $\rho_{PANC} = 30\%$ with PANC and $\rho_{NL} = 15\%$ without PANC.

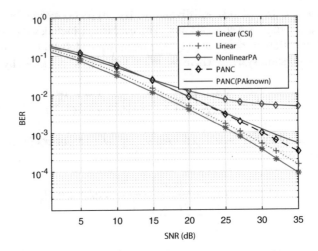

Fig. 4.8 BER versus SNR for PANC with model estimation in an OFDM system with 16-QAM modulation on $N = 512$ subcarriers, for the case of linear PA and NL PA. A Wiener–Hammerstein type nonlinear PA was considered where $g[\cdot]$ emulates an SSPA model with $p = 2$ with clipping level $\gamma = 3$ dB. Curves for linear PA with perfect channel state information (linear (CSI)), and PANC assuming perfect acknowledge of PA model (PANC (PA known)), are included for comparison purposes

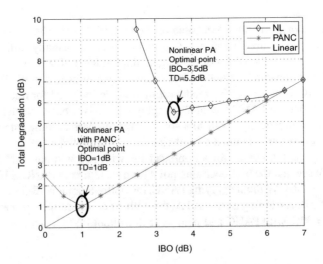

Fig. 4.9 Total degradation versus IBO for PANC with model estimation in an OFDM system with 16-QAM modulation on $N = 512$ subcarriers, for the case of linear PA and nonlinear PA. Wiener–Hammerstein type nonlinear PA was considered where $g[\cdot]$ was an SSPA model with $p = 2$ and clipping level $\gamma = 4$ dB

4.6   A Case of Study: Linearization of Class AB and Envelope Tracking PAs

Highly nonlinear power amplifier (PA) exhibits different characteristics depending on the input power regions, making difficult to compensate the distortion with conventional digital predistorters architectures.

To overcome these problems, the use of a different approach for the PD design based on a partitioned digital predistorter (P-DPD) can be evaluated. The proposed partitioned PD decomposes the entire range of the transfer function in separated parts. Each part presents a weakly nonlinear transfer that can be compensated using a low order PD. In addition to complexity reduction the advantages of the partitioned PD are (due to low order DPD per partition) better numerical properties associated with coefficient estimation and relaxed requirements in terms of sampling frequency.

The number of partitions and the threshold definition depend on the nonlinearity shape and the probability density function (pdf) of the input signal. The trade-off between the number of partitions, polynomial order, and estimation accuracy is taken into account in the definition of the partitions.

4.6.1   Partitioned Predistorter

The proposed predistorter is based on the memory polynomial (MP) model [11]. That model has been employed in predistortion techniques showing a very good performance [69]. The main characteristics of the MP model, which we exploit regarding real-time applications, are its modularity and simplicity. The output of a MP predistorter can be written as:

$$d(n) = \sum_{p=0}^{P-1} \sum_{k=0}^{M-1} \theta_{pk}(n)\psi_p^*(n-k), \qquad (4.20)$$

where the basis function of the MPs are defined by $\psi_p(n) = y(n)|y(n)|^{2p}$. P and M are the polynomial order and memory depth, respectively. The election of polynomial order and memory depth is a function of the nonlinearity shape and memory effects. For weakly nonlinear PA (class A amplifiers) good performance can be reached with low polynomial order (5th or 7th). However, for highly nonlinear PAs (class AB, class B amplifiers), a large polynomial order is required. In this case, the DPD will suffer from numerical implementation drawbacks when implemented in DSP hardware with finite resolution. Moreover, the sampling requirements are more stringent. These drawbacks motivate the implementation of a partitioned digital predistorter (P-DPD) by decomposing the entire range of the transfer function in separated parts which present a weakly nonlinearity that can be compensated with individual low order models.

The signal threshold decomposition is based on [70] following the same structure presented in [71]. Since the model is a polynomial, we consider that the optimal number of sectors decomposition would be related to the behavior of the third harmonic component for this class AB PA as described in [72, 73]. So the optimum number of sectors could be considered in terms of the above and the hardware resolution.

The signal threshold decomposition can be implemented using a soft limiter settled with different thresholds $U = [U_1, U_2, \ldots, U_Q]$:

$$
x_q(n) = \begin{cases}
0 & \text{if} |x(n)| \leq U_{q-1}, \\
(|x(n)| - U_{q-1})e^{j\phi} & \text{if } U_{q-1} < |x(n)| \leq U_q, \\
(U_q - U_{q-1})e^{j\phi} & \text{if } |x(n)| > U_q
\end{cases}
$$

where ϕ is the phase of $x(n)$. An example of signal decomposition with three threshold levels U_1, U_2, and U_3 is illustrated in Fig. 4.10a.

Following the signal decomposition approach developed in [71], the output of the P-DPD $u(n)$ can be obtained then as follow:

$$
u(n) = \begin{cases}
F_1[x(n)] & \text{if } 0 \leq |x(n)| \leq U_1 \\
F_2[x(n)] & \text{if } U_1 \leq |x(n)| \leq U_2, \\
\ldots \quad \ldots \\
F_q[x(n)] & \text{if } U_{q-1} \leq |\tilde{x}(n)| \leq U_q
\end{cases}
$$

where $F_i[.]$ is the nonlinear transfer function for the signal whose magnitude falls in the interval (U_{i-1}, U_i). In our implementation, the fitting functions $F_i[.]$ are polynomials. The input samples are processed separately using different polynomial functions and finally added to generate the P-DPD output as is illustrated in Fig. 4.10b.

4.6.1.1 Partitioned DPD Parameter Estimation

The P-DPD coefficients are estimated using an indirect learning structure. Input and output signals of the PA are fed to the indirect learning filter such that the post-inverse of the PA model is identified. Then the parameters are directly copied to the DPD block, see e.g., [69, 74].

Using the vector decomposition, the PA output signal can be expressed as:

$$
y = \sum_{q=1}^{Q} y_q(n), \tag{4.21}
$$

where y_q is the output of the q-branch of the signal threshold decomposition block. P-DPD output signal can be expressed as:

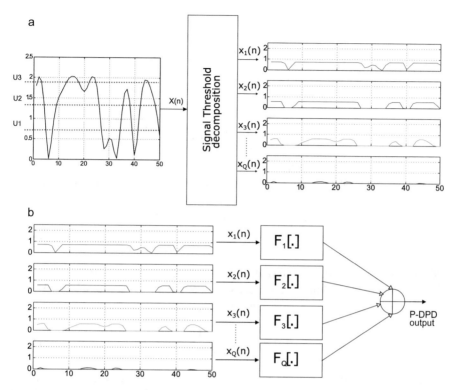

Fig. 4.10 (**a**) Vector decomposition. (**b**) Vector composition

$$d = \mathbf{\Psi}^H \boldsymbol{\theta}, \tag{4.22}$$

where $\boldsymbol{d}$ is $(N \times 1)$ vector representing the N samples of the P-DPD output, $\boldsymbol{\theta}$ denotes the predistorter coefficients vector of length $L = \sum_{q=1}^{Q} M_q P_q$, where M_q and P_q are the memory depth of the polynomial order, respectively, employed for each partition, and

$$\mathbf{\Psi} = \begin{bmatrix} \boldsymbol{\phi}(0) \ \boldsymbol{\phi}(1) \cdots \boldsymbol{\phi}(N-1) \end{bmatrix} \tag{4.23}$$

is an $(L) \times N$ matrix formed by the basis function defined by

$$\boldsymbol{\phi}(n) = [\boldsymbol{\psi}^{1^T}(n) \cdots \boldsymbol{\psi}^{1^T}(n - M_1 + 1), \ \boldsymbol{\psi}^{2^T}(n) \cdots \boldsymbol{\psi}^{2^T}(n - M_2 + 1), \cdots,$$

$$\boldsymbol{\psi}^{Q^T}(n) \cdots \boldsymbol{\psi}^{Q^T}(n - M_Q + 1)], \tag{4.24}$$

where

$$\boldsymbol{\psi}^1(n) = [\psi_{1_0}(n) \cdots \psi_{1_{P_1-1}}(n)]^T$$

$$\boldsymbol{\psi}^2(n) = [\psi_{2_0}(n) \cdots \psi_{2_{P_2-1}}(n)]^T$$

$$\cdots$$

$$\boldsymbol{\psi}^Q(n) = [\psi_{Q_0}(n) \cdots \psi_{Q_{P_Q-1}}(n)]^T,$$

(4.25)

where $\psi_{p_q}(n) = y_q(n)|y_q(n)|^p$.

The desired P-DPD output is the original transmitted signal x and is given

$$x = [x(0)\; x(1) \cdots x(N-1)]^T,$$

(4.26)

where x is a $(N \times 1)$ vector representing the N samples of the desired P-DPD output. A least-squares algorithm is implemented to obtain the P-DPD coefficients. The LS solution for (4.22) replacing $d = x$ is given by [75]

$$\theta = \left(\boldsymbol{\Psi}\boldsymbol{\Psi}^H\right)^{-1}\boldsymbol{\Psi}x.$$

(4.27)

In this estimator, the measurement noise affects the data matrix $\boldsymbol{\Psi}$, while in the ordinary LS solution, the measurement noise lies in the observation vector x. In this case, the estimator defined by Eq. (4.27) is called *data least squares*.

4.6.2 Partitions Allocations

Considering a typical amplifier, its dynamic range can be divided into three principal regions: (a) cutoff region, that is the region placed close to zero value and its transfer function is highly nonlinear for class AB and class B amplifiers, (b) linear region, where the response is almost linear, partitions are not required and can be handled with a low polynomial order, (c) gain compression region, the region is extended from the medium-large input signal level until saturation. The distortion generated in this region is important and need to be compensated.

The partitions can be uniformly or nonuniformly distributed around the PA input dynamic range.

Assuming that the input signal envelope is defined in the range of [0, x_{max}], in case of uniform partitions this interval is divided into M equally sized intervals given by $\Delta = \frac{x_{max}}{M}$.

This uniform-distribution approach is particularly attractive because of its computational simplicity. However, due to the shape of the PA nonlinearity and the distribution of the input signal, we can infer that the regions with high nonlinear response and with large occurrence probability (due to OFDM symbols distribution) will require more partitions than regions with linear or weakly nonlinear response and with low occurrence probability. If the partition size is variable (nonuniform partitions), it enables the model to fit hard nonlinearities accurately using a moderate number of partitions.

4.6.2.1 Input Distribution-Based Partition Allocation

The partitions can be allocated to match the distribution of the input signal to be transmitted through the nonlinear PA. A key modulation signal to consider in our work is OFDM. With that objective we follow a similar approach to optimize nonuniform quantizers.

Considering an OFDM system with large number of subcarriers ($N > 32$), PA input signal probability density function (pdf) can be well approximated by a complex Gaussian random process, i.e., its envelope has a Rayleigh distribution, that can be written as:

$$f_R(x) = \frac{x}{\sigma^2} \exp\left(-\frac{x^2}{2\sigma^2}\right), \tag{4.28}$$

where σ^2 denotes the variance of the signal.

Definition Let X be a continuous random variable with generic pdf $f(x)$ defined in the range $c_1 < x < c_2$. And, let $Y = g(X)$ be an invertible function of X with inverse function $X = v(Y)$. Then, the *pdf* of y is given by

$$f_y(y) = f_x(v(y)) \left|v'(y)\right|. \tag{4.29}$$

The PA response can be modeled with a K-th order polynomial function given by

$$g_{pa}(n) = \sum_{k=0}^{K-1} b_k x(n)^k, \tag{4.30}$$

where b_k are the polynomial coefficients employed to model the PA response. The PA response for a typical PA is plotted in Fig. 4.11a considering different bias. The P-th order inverse polynomial of this model can be expressed as:

$$z_I(n) = \sum_{p=0}^{P-1} c_p y(n)^p, \tag{4.31}$$

where c_p are the polynomial coefficients that model the inverse of the PA response. Then, following the analytical expression defined by Eq. (4.29), the *pdf* of the PA output data can be expressed as:

Fig. 4.11 (**a**) Typical power
amplifier AM-AM response
with different bias. (**b**)
Probability density
function—PA output

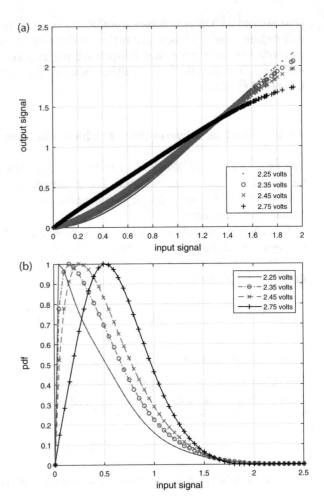

$$f_y(y) = f_x\left(\sum_{p=0}^{P} c_p y(n)^p\right)\left|\sum_{p=0}^{P} pc_p y(n)^{p-1}\right|$$

$$= \frac{\sum_{p=0}^{P} c_p y(n)^p}{\sigma^2}\exp\left(-\frac{(\sum_{p=0}^{P} c_p y(n)^p)^2}{2\sigma^2}\right)\left|\sum_{p=0}^{P} pc_p y(n)^{p-1}\right| (4.32)$$

Figure 4.11 illustrates the output of a class AB power amplifier with different
bias. From this figure, we can infer that when the bias is reduced (large power
efficiency), the output signal distribution is concentrated near to the cutoff region.
For large bias, the maximum of the *pdf* is moved to the linear region.

The input distribution-based partition will make the interval with large probability smaller, allowing large distortion in regions of lower probability. The input

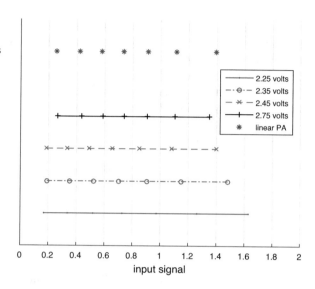

Fig. 4.12 Nonuniform partitions allocation for typical PA with different bias

distribution-based approach makes the intervals smaller in those regions that have more probability mass. In our application, the intervals near to the origin will be smaller than intervals in the linear region and close to the saturation zone.

Assuming the knowledge of the source distribution (PA output), the Lloyd-Max algorithm [76] can be employed to calculate the intervals that define each partition The centroid condition is expressed as:

$$s_j = \frac{\int_{y_j}^{y_{j+1}} y f_y(y)}{\int_{y_j}^{y_{j+1}} f_y(y)}, \tag{4.33}$$

where s_j the quantized output level. The decision boundary j is the midpoint of two neighboring quantized levels, given by

$$U_j = \frac{y_{j+1} + y_j}{2}. \tag{4.34}$$

To obtain the decision boundaries, (4.33) and (4.34) can be solved iteratively. The obtained vector $U_{pdf} = [U_1, U_2, \ldots, U_Q]$, defines the partitions allocated following the pdf distribution of the PA output signal. The allocation results are plotted in Fig. 4.12 for different PA bias.

4.6.2.2 μ Law Partitions Allocation

The implementation of the input distribution-based algorithm requires a large computational complexity that cannot be afforded in the majority of the wireless applications. Following the idea of nonuniform partition allocation, we propose the allocation of the partitions using a logarithmic function, where the low-amplitude regions are emphasized.

By simulations, we verify that allocating the partitions using a μ-law compander [77] allows to obtain promising results. Considering an input signal with a dynamic range of $[0, x_{\max}]$, initially the interval is divided in M equal-size intervals $\Delta = \frac{x_{\max}}{M}$. The partitions vector can be written as $U = [0, \Delta, 2\Delta, M-1\Delta]$. Each element of this vector is mapped by a μ-law compander function $g(\cdot)$ given by

$$U_\mu = g_\mu(U) = \frac{U_{\max}}{\mu} \left((1+\mu)^{\frac{|U|}{U_{\max}}} - 1\right), \qquad (4.35)$$

where U_μ is the partitions vector with nonuniform μ distribution. Low implementation complexity makes this allocation technique the preferred option.

4.6.2.3 Iterative Partitions Allocation Technique

The use of μ-law partition allocation reaches very good results with low implementation complexity. However, this allocation is ad-hoc, and is not optimized for a particular PA response. In order to improve the P-DPD performance, we proposed a iterative allocation technique that is optimized to minimize a function of the linearization error. The algorithm is initialized allocating the partitions using a μ-law distribution. With the data decomposed using this partition, the P-DPD coefficients are obtained with a set of N PA input/output data. With the obtained coefficients, the initial linearization error is calculated.

Initially, the first partition is moved one step in the positive direction (increasing its value), a new set of coefficients is calculated and the linearization error is obtained. This error is compared with the previous error. If the difference is less than a tolerance threshold, the optimization of this threshold level is stopped. Otherwise, the threshold level is adjusted in the direction indicated by the sign of the difference. The procedure is repeated until the tolerance threshold level is reached. The same procedure is applied to obtain the allocation of each partition.

The allocation technique is summarized in Table 4.2. An example of partitions allocation using μ law allocation, iterative allocation, and *pdf*-optimized allocation is depicted in Fig. 4.13.

4.6.3 Numerical Evaluation

The performance of partitioned and conventional DPD is evaluated considering an envelope-tracking PA. The amplitude-to-amplitude (AM/AM) and amplitude-to-phase (AM/PM) characteristics for different drain supply voltages V_{dd} for an envelope-tracking PA are given by [78, 79]

Table 4.2 Iterative partition allocation algorithm

Initialization:

$X = [x(1), x(2), \cdots, x(N)]$,   Input sequence

$Y = [y(1), y(2), \cdots, y(N)]$ PA Output sequence

$\lambda_0 = \lambda_\mu$, $\boldsymbol{\theta} = \left(\boldsymbol{\Psi}\boldsymbol{\Psi}^H\right)^{-1}\boldsymbol{\Psi}x$

$\rho = constant$,   tolerance threshold (termination condition)

$step = [step(1), step(2), \cdots, step(Q)]$,   define the step to optimize each the allocation of each partition

FOR $q = 1$ to Q

{

   i=0, flag=1, ini=1

   DO WHILE $b \neq 0$,   execute until the minimum error is obtained

   {

   i=i+1

   $\lambda(q) = \lambda(q) + step(q) \times flag$

   Coefficient estimation using the updated partition: $\boldsymbol{\theta}$

   Linearization error: $|e(i)|^2 = \left(\frac{\sum_{n=1}^{K} |x(n) - \boldsymbol{\Psi}(n)\boldsymbol{\theta}|^2}{\sum_{n=1}^{K} |x(n)|^2} \right)$

   IF $d = |e(i)|^2 - |e(i-1)|^2 < 0$

   Verify the termination condition:

     IF $d < \rho$

     $b = 0$ $\lambda(q)$ reached the optimal value

     ELSE

     Continue the search in the same direction $flag = 1$

     END

   ELSE

   Modify the search direction $flag = -1$

   }

}

$$F_{AM/AM} = \frac{(g_0 + g_1 A)A}{\sqrt{(1 + [(g_0 + g_1 A)A]^s)}}$$

$$F_{AM/PM} = \arctan(a \exp^{bA}(A - p)^2 + q),$$

where the parameters g_0, g_1, L, s, a, b, p, and q depend on V_{dd} and are detailed in [78, 79].

The performance of P-DPD and conventional DPD structures is evaluated. We assume the same memory depth for each P-DPD branch and the single DPD, i.e., $M_1 = M_2 \ldots M_q = M_s$. Third-order polynomials are employed for each P-DPD branch.

Fig. 4.13 Partitions
allocation using; μ law
allocation, iterative
allocation, and *pdf*-optimized
allocation

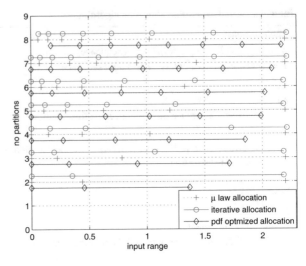

Fig. 4.14 Gain and AM-PM
curves for linearized ET PA.
The PPD is implemented with
4 partitions (third-order
polynomial) and $M_q = 3$, and
the PD employ a ninth-order
polynomial and memory
depth $M = 3$

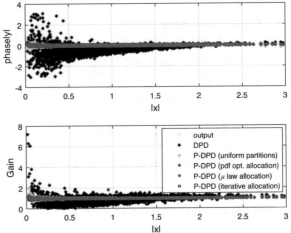

Gain and AM/PM curves for both predistorter implementations are plotted in
Fig. 4.14. Spectral regrowth curves are shown in Fig. 4.15 for P-DPD and DPD
with different polynomial and memory depth. The figure shows the power spectral
density (PSD) of the power amplifier input and output signals with and without
predistortion. The P-DPD is implemented using fourth partitions with a memory
depth 3 and a third-order polynomial. The DPD is implemented using a ninth-order
polynomial with a memory depth of 3. The performance has been evaluated, in terms
of ACPR and EVM, considering a P-DPD with different number of partitions. ACPR
and EVM results as a function of the number of partitions are shown in Figs. 4.16
and 4.17, respectively.

Fig. 4.15 Power spectral density of the linearized PA output

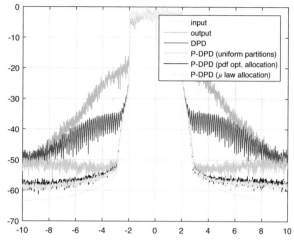

Fig. 4.16 ACPR in function of number of partitions for P-DPD. Conventional DPD is included for comparison purpose

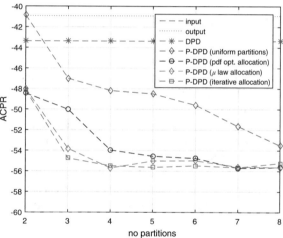

From these results, it can be observed that the partitioned structure outperforms the conventional techniques. In-band and out-of-band results show that by using only four partitions, the P-DPD reaches very good results with moderated implementation complexity. Implementation complexity of the digital block and optimization of the partitions in function of the type of the PA are challenging issues that must be further evaluated.

Fig. 4.17 EVM in function of the number of partitions. Each partition of the P-DPD is modeled with a third-order polynomial and $M_q = 3$, and the PD employ a ninth-order polynomial and memory depth $M = 3$

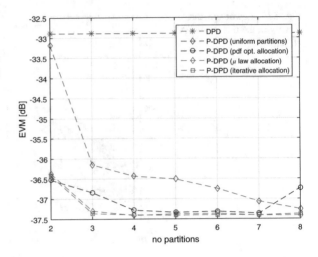

4.7 Summary of the Key Points

- The power amplifier is the main power consumer in typical wireless transceiver. The improvements in terms of energy efficiency can be obtained by using nonlinear distortion compensation techniques that can be applied either in the receiver or the transmitter side.
- Transmitter side compensation techniques as PAPR reduction algorithms and digital predistorters are mainly applied in base stations where more computational resources are available.
- Receiver-side compensation methods are good candidates for the uplink of IoT systems where reduced computational capacity and energy are available. The more complex processing task is moved to the base stations.

References

1. P.M. Lavrador, T.R. Cunha, P.M. Cabral, J.C. Pedro, The linearity-efficiency compromise. IEEE Microw. Mag. **11**(5), 44–58 (2010)
2. P. Kenington, *High-Linearity RF Amplifier Design* (Artech House, Norwood, 2000)
3. S. Cripps, *Advanced Techniques in RF Power Amplifiers Design* (Artech, Norwood, 2002)
4. F. Giannini, E. Limiti, P. Colantonio, *High Efficiency RF & Microwave Solid State Power Amplifiers* (John Wiley & Sons, Hoboken, 2009)
5. B. Berglund, T Nygren, K. Sahlman, RF multicarrier amplifier for third generation systems. Ericsson Rev. **78**(4), 184–189 (2001)
6. P.K. Singya, N. Kumar, V. Bhatia, Mitigating NLD for wireless networks: effect of nonlinear power amplifiers on future wireless communication networks. IEEE Microw. Mag. **18**(5), 73–90 (2017)

7. J. Joung, C.K. Ho, K. Adachi, S. Sun, A survey on power-amplifier-centric techniques for spectrum- and energy-efficient wireless communications. IEEE Commun. Surv. Tutorials **17**(1), 315–333 (2015). First quarter
8. A. Katz, J. Wood, D. Chokola, The evolution of PA linearization: from classic feedforward and feedback through analog and digital predistortion. IEEE Microw. Mag. **17**(2), 32–40 (2016)
9. G. Wunder, R.F.H. Fischer, H. Boche, S. Litsyn, J. No, The PAPR problem in OFDM transmission: new directions for a long-lasting problem. IEEE Signal Process. Mag. **30**(6), 130–144 (2013)
10. J. Kim, K. Konstantinou, Digital predistortion of wideband signals based on power amplifier model with memory. Electron. Lett. **37**(23), 1417–1418 (2001)
11. D. Morgan, Z. Ma, J. Kim, M. Zierdt, J. Pastalan, A generalized memory polynomial model for digital predistortion of RF power amplifiers. IEEE Trans. Signal Process. **54**(10), 3852–3860 (2006)
12. P. Jaraut, M. Rawat, P. Roblin, Digital predistortion technique for low resource consumption using carrier aggregated 4G/5G signals. IET Microwaves Antennas Propag. **13**(2), 197–207 (2019)
13. O. Hammi, S. Buomaiza, F. Ghannouchi, On the robustness of digital predistortion function synthesis and average power tracking for highly nonlinear power amplifiers. IEEE Trans. Microwave Theory Tech. **55**(6), 1382–1389 (2007)
14. S. Mann, M. Beach, P. Warr, J. McGeehan, Increasing the talk-time of mobile radios with efficient linear transmitter architectures. Electron. Commun. Eng. J. **13**(2), 65–76 (2001)
15. E. Aschbacher, M. Rupp, Modeling and identification of a nonlinear power-amplifier with memory for nonlinear digital adaptive pre-distortion, in *Proceedings of IEEE Signal Processing Advances in Wireless Communications, SPAWC 2003*, Rome, vol. 1 (2003), pp. 658–662
16. L. Guan, A. Zhu, Green communications: digital predistortion for wideband RF power amplifiers. IEEE Microw. Mag. **15**(7), 84–99 (2014)
17. C. Eun, E.J. Powers, A new Volterra predistorter based on the indirect learning architecture. IEEE Trans. Signal Process. **45**(1), 223–227 (1997)
18. L. Ding, G.T. Zhou, D.R. Morgan, Z. Ma, J.S. Kenney, J. Kim, C.R. Giardina, A robust digital baseband predistorter constructed using memory polynomials. IEEE Trans. Commun. **52**(1), 159–165 (2004)
19. P. Gilabert, G. Montoro, E. Bertran, On the Wiener and Hammerstein models for power amplifier predistortion, in *Proceedings of Asia-Pacific Microwave Conference (APMC)* (2005)
20. F. Gregorio, S. Werner, J. Cousseau, T. Laakso, Receiver cancellation technique for nonlinear power amplifier distortion in SDMA-OFDM systems. IEEE Trans. Vehic. Tech. **56**(5) Part I, 2499–2516 (2007)
21. F. Gregorio, Analysis and compensation of nonlinear power amplifier effects in multi-antenna OFDM systems, Ph.D. Thesis, Helsinki University of Technology (2007)
22. E. Olfat, M. Bengtsson, Joint channel and clipping level estimation for OFDM in IoT-based networks. IEEE Trans. Signal Process. **65**(18), 4902–4911 (2017)
23. L.D. Quach, S.P. Stapleton, A postdistortion receiver for mobile communications. IEEE Trans. Veh. Technol. **42**(4), 604–616 (1993)
24. S.P. Stapleton, L. Quach, Reduction of adjacent channel interference using postdistortion, in *Proceedings of IEEE Vehicular Technology Conference, VTC'02-Fall*, Vancouver, vol. 2 (1992), pp. 915–918
25. C.H. Lee, V. Postoyalko, T. O'Farrell, Enhanced performance of ROF link for cellular mobile systems using postdistortion compensation, in *Proceedings of IEEE Personal, Indoor and Mobile Radio Communications, PIMRC 2004*, Barcelona, vol. 4 (2004), pp. 2772–2776
26. J. Macdonald, Nonlinear distortion reduction by complementary distortion. IRE Trans. Audio **7**(5), 128–133 (1959)
27. M. Aziz, M.V. Amiri, M. Helaoui, F.M. Ghannouchi, Statistics-based approach for blind post-compensation of modulator's imperfections and power amplifier nonlinearity. IEEE Trans. Circuits Syst. Regul. Pap. **66**(3), 1063–1075 (2019)

28. D.D. Falconer, Adaptive equalization of channel nonlinearities in QAM data transmission systems. Bell Syst. Tech. **57**, 2589–2611 (1978)
29. S. Benedetto, E. Biglieri, Nonlinear equalization of digital satellite channels. IEEE J. Sel. Areas Commun. **1**(1), 57–62 (1983)
30. S. Benedetto, A. Gersho, R.D. Gitlin, T.L. Lim, Adaptive cancellation of nonlinear intersymbol interference for voiceband data transmission. IEEE J. Sel. Areas Commun. **2**(5), 765–777 (1984)
31. D. Kim, G.L. Stuber, Clipping noise mitigation for OFDM by decision-aided reconstruction. IEEE Commun. Lett. **3**(1), 4–6 (1999)
32. J. Tellado, L.M.C. Hoo, J.M. Cioffi, Maximum-likelihood detection of nonlinearly distorted multicarrier symbols by iterative decoding. IEEE Trans. Commun. **51**, 218–228 (2003)
33. Y. Xiao, S. Li, X. Lei, Y. Tang, Clipping noise mitigation for channel estimation in OFDM systems. IEEE Commun. Lett. **10**(6), 474–476 (2006)
34. H. Chen, A.M. Haimovich, Iterative estimation and cancellation of clipping noise for OFDM signals. IEEE Commun. Lett. **7**(7), 305–307 (2003)
35. R. AliHemmati, P. Azmi, Iterative reconstruction-based method for clipping noise suppression in OFDM systems. *IEE Commun. Proc.* **152**(4), 452–456 (2005)
36. S.V. Zhidkov, Receiver synthesis for nonlinearly amplified OFDM signal, in *Proceedings of IEEE Workshop on Signal Processing Systems, SIPS 2003*, Seoul (2003), pp. 387–392
37. C.A.R. Fernandes, J.C.M. Mota, G. Favier, Analysis and power diversity-based cancellation of nonlinear distortions in OFDM systems. *IEEE Trans. Signal Process.* **60**(7), 3520–3531 (2012)
38. V. del Razo, T. Riihonen, F. Gregorio, S. Werner, R. Wichman, Nonlinear amplifier distortion in cooperative amplify-and-forward OFDM systems, in *2009 IEEE Wireless Communications and Networking Conference* (2009), pp. 1–5
39. H. Bouhadda, R. Zayani, H. Shaiek, D. Roviras, R. Bouallegue, Iterative receiver cancellation of nonlinear power amplifier distortion in FBMC/OQAM system, in *2015 IEEE 11th International Conference on Wireless and Mobile Computing, Networking and Communications (WiMob)* (2015), pp. 691–695
40. A.S. Tehrani, H. Cao, A. Behravan, T. Eriksson, C. Fager, Successive cancellation of power amplifier distortion for multiuser detection, in *2010 IEEE 72nd Vehicular Technology Conference – Fall* (2010), pp. 1–5
41. Z. Alina, O. Amrani, On digital post-distortion techniques. *IEEE Trans. Signal Process.* **64**(3), 603–614 (2016)
42. F.H. Gregorio, S. Werner, J. Cousseau, J. Figueroa, R. Wichman, Receiver-side nonlinearities mitigation using an extended iterative decision-based technique. Signal Proc. **91**(8), 2042–2056 (2011)
43. J.K. Cavers, Amplifier linearization using a digital predistorter with fast adaptation and low memory requirements. IEEE Trans. Veh. Technol. **(39)**(4), 374–382 (2000)
44. A.A.M. Saleh, J. Salz, Adaptive linearization of power amplifiers in digital radio systems. Bell Syst. Tech. J. **62**(4), 1019–1033 (1983)
45. A.N. D'Andrea, V. Lottici, R. Reggiannini, Nonlinear predistortion of OFDM signals over frequency-selective fading channels. IEEE Trans. Commun. **49**(5), 837–843 (2001)
46. M. Faulkner, M. Johansson, Adaptive linearization using predistortion-experimental results. IEEE Trans. Veh. Technol. **43**(2), 323–332 (1994)
47. M.Y. Cheong, S. Werner, M.J. Bruno, J.L. Figueroa, J.E. Cousseau, R. Wichman, Adaptive piecewise linear predistorters for nonlinear power amplifiers with memory. IEEE Trans. Circuits Syst. Regul. Pap. **59**(7), 1519–1532 (2012)
48. J.E. Cousseau, J.L. Figueroa, S. Werner, T.I. Laakso, Efficient nonlinear wiener model identification using a complex-valued simplicial canonical piecewise linear filter. IEEE Trans. Signal Process. **55**(5), 1780–1792 (2007)
49. Y. Guo, J.R. Cavallaro, Enhanced power efficiency of mobile OFDM radio using predistortion and post-compensation, in *Proceedings of IEEE Vehicular Technology Conference, VTC'02-Fall*, Vancouver, vol. 1 (2002), pp. 214–218

50. M.Y. Cheong, S. Werner, T.I. Laakso, Design of predistorters for power amplifiers in future mobile communications systems, in *Nordic Signal Processing Symposium NORSIG'04*, Espoo (2004)
51. H.W. Kang, Y.S. Cho, D.H. Youn, On compensating nonlinear distortions of an OFDM system using an efficient adaptive predistorter. IEEE Trans. Commun. **47**(4), 522–526 (1999)
52. M. Schetzen, *The Volterra and Wiener Theories of Nonlinear Systems* (J. Wiley Sons, Hoboken, 1980)
53. K.F. To, P.C. Ching, K.M. Wong, Compensation of amplifier nonlinearities on wavelet packet division multiplexing, in *Proceedings of IEEE International Conference Acoustics, Speech, Signal Processing, ICASSP'01*, Utah, vol. 4 (2001), pp. 2669–2672
54. J. Tsimbinos, K.V. Lever, Nonlinear system compensation based on orthogonal polynomial inverses. IEEE Trans. Circuits Syst. I: Fundam. Theory Appl. **48**(4), 406–417 (2006)
55. C.H. Tseng, E.J. Powers, Nonlinear channel equalization in digital satellite systems, in *Proceedings of IEEE Global Telecommunications Conference, GLOBECOM 1993*, Houston, vol. 1 (1993), pp. 1639–1643
56. E. Biglieri, S. Barberis, M. Catena, Analysis and compensation of nonlinearities in digital transmission systems. *IEEE J. Sel. Areas Commun.* **6**(1), 42–51 (1988)
57. H.W. Kang, Y.S. Cho, D.H. Youn, Adaptive precompensation of Wiener systems. IEEE Trans. Signal Process. **46**(10), 2825–2829 (1998)
58. T. Liu, S. Boumaiza, F.M. Ghannouchi, Pre-compensation for the dynamic nonlinearity of wideband wireless transmitters using augmented Wiener predistorters, in *Proceedings of Asia-Pacific Microwave Conference, APMC 2005*, Suzhou (2005)
59. T. Liu, S. Boumaiza, F.M. Ghannouchi, Identification and pre-compensation of the electrical memory effects in wireless transceivers, in *Proceedings of IEEE Radio and Wireless Symposium*, San Diego (2006), pp. 535–538
60. R. Raich, H. Qian, G. T. Zhou, Orthogonal polynomials for power amplifier modeling and predistorter design. IEEE Trans. Veh. Technol. **53**(5), 1468–1479 (2004)
61. J. Kim, K. Konstantinou, Digital predistortion of wideband signals based on power amplifier model with memory. IEE Electronics Lett. **37**, 1417–1418 (2001)
62. Y. Liu, J. Zhou, W. Chen, B. Zhou, A robust augmented complexity-reduced generalized memory polynomial for wideband RF power amplifiers. IEEE Trans. Ind. Electron. **61**(5), 2389–2401 (2014)
63. N. Kelly, W. Cao, A. Zhu, Preparing linearity and efficiency for 5G: digital predistortion for dual-band Doherty power amplifiers with mixed-mode carrier aggregation. IEEE Microw. Mag. **18**(1), 76–84 (2017)
64. S. Zhang, W. Chen, Z. Feng, Low sampling rate digital predistortion of power amplifier assisted by bandpass RF filter, in *2012 Asia Pacific Microwave Conference Proceedings* (2012), pp. 962–964
65. C. Nader, W. Van Moer, K. Barbe, N. Bjorsell, P. Handel, Harmonic sampling and reconstruction of wideband undersampled waveforms: breaking the code. IEEE Trans. Microwave Theory Tech. **59**(11), 2961–2969 (2011)
66. Q. Zhang, Y. Liu, J. Zhou, W. Chen, A complexity-reduced band-limited memory polynomial behavioral model for wideband power amplifier, in *2015 IEEE International Wireless Symposium (IWS 2015)* (2015), pp. 1–4
67. D. Dardari, V. Tralli, A. Vaccari, A theoretical characterization of nonlinear distortion effects in OFDM systems. IEEE Trans. Commun. **48**(10), 1755–1764 (2000)
68. A.A.M. Saleh, Frequency-independent and frequency-dependent nonlinear models of TWT amplifiers. IEEE Trans. Commun. **29**(11), 1715–1720 (1981)
69. L. Ding, G. Zhou, D. Morgan, Z. Ma, J. Kenney, J. Kim, C. Giardina, A robust digital baseband predistorter constructed using memory polynomials. IEEE Trans. Commun. **52**(1), 159–165 (2004)
70. K. Gharaibeh, Behavioral modeling of nonlinear power amplifiers using threshold decomposition based piecewise linear approximation. IEEE Radio Wirel. Symp. 755–758 (2008)

71. A. Zhu, P. Draxler, C. Hsia, T. Brazil, D. Kimball, P. Asbeck, Digital pre distortion for envelope tracking power amplifiers using decomposed piecewise Volterra series. IEEE Trans. Microwave Theory Tech. **56**(10), 2237–2247 (2008)
72. C. Fager, J.C. Pedro, N. Borges de Carvalho, H. Zirath, Prediction of IMD in LDMOS transistor amplifiers using a new large signal model. IEEE Trans. MTT **50**(12), 2834–2842 (2002)
73. C. Fager, J.C. Pedro, N. Borges de Carvalho, H. Zirath, Intermodulation distortion behavior in LDMOS transistor amplifiers. IEEE MTT Digest. 131–134 (2002)
74. C. Eun, E.J. Powers, A predistorter design for a memory-less nonlinearity preceded by a dynamic linear system, in *IEEE GLOBECOM 95*, vol. 1 (1995), pp. 152–156
75. G. Golub, C.F. Van Loan, *Matrix Computations* (The Johns Hopkins University Press, Baltimore, 1993)
76. A.V. Trushkin, On the design of an optimal quantizer. IEEE Trans. Inf. Theory **39**(4), 1180–1194 (1993)
77. C. Hsu, H. Liao, PAPR reduction using the combination of precoding and Mu-Law companding techniques for OFDM systems, in *2012 IEEE 11th International Conference on Signal Processing*, vol. 1 (2012), pp. 1–4
78. A. Hekkala, A. Kotelba, M. Lasanen, Compensation of linear and nonlinear distortions in envelope tracking amplifier, in *IEEE 19th International Symposium on Personal, Indoor and Mobile Radio Communications, 2008. PIMRC 2008* (2008), pp. 1–5
79. A. Kotelba, A. Hekkala, M. Lasanen, Compensation of time misalignment between input signals in envelope-tracking amplifiers, in *IEEE 19th International Symposium on Personal, Indoor and Mobile Radio Communications, 2008. PIMRC 2008* (2008), pp. 1–5

Chapter 5
ADC in Broadband Communications

Abstract Efficient and application-oriented analog-to-digital conversion (ADC) plays a key role on the performance of any communication system. Among the different available architectures, there exists a trade-off between sampling rate and resolution, both also related to the power consumption of the device. In addition, nonlinear distortion can severely reduce the digital dynamic range of the converted signal, thus reducing the effective resolution with the corresponding negative effect on the receiver sensitivity. In this sense, the selection and development of accurate models and compensation strategies are required to restore adequate performance. For example, the complexity of the models and compensation algorithms must also be considered in order to achieve an efficient solution. While ADCs used to sample narrowband signals have little memory effects and allow for simple models and compensation techniques, sampling of broadband signals introduces longer memory effects and more complex nonlinear dynamic models are required (Volterra, piece-wise linear models). Finally, adequate ADC performance metrics and figures of merit have to be carefully chosen to evaluate the quality of the compensation for the application at hand, as well as the measurement set-up and validation tests. In this chapter, we describe several of the available ADC architectures in terms of the achievable resolution and sampling rate, and the trade-off between them. Narrowband as well as wideband modeling and compensation techniques are described and proposed, depending on the particular ADC and the application at hand. Measurement related issues are also discussed.

5.1 ADC Architectures

In general, the selection of a particular ADC architecture is dependent on the requirements of the system to be sampled. While each ADC architecture has its own advantages and limitations when compared to the others, and there is usually a trade-off between sampling speed and resolution difficult to overcome [1], some strategies are also available to combine several ADCs into more complex sampling structures to improve their characteristics [2]. In the following sections the main

© Springer Nature Switzerland AG 2020
F. Gregorio et al., *Signal Processing Techniques for Power Efficient Wireless Communication Systems*, Signals and Communication Technology,
https://doi.org/10.1007/978-3-030-32437-7_5

basic ADC architectures and structures will be introduced and briefly described. Although implementation and technological issues are out of the scope of this book, interested readers are encouraged to surf through the included references.

5.1.1 Flash: High Conversion Speed, Low Resolution

An ADC flash of b bits is an array of 2^b comparators which are connected on one terminal to the analog input signal to be sampled, and on the other terminal to a set of reference voltages determined by a resistors ladder voltage divider, as shown in Fig. 5.1. Then, for a particular amplitude value of the input signal, all the comparators with lower reference voltage will be triggered in a *thermometric* manner, which can be effectively translated to binary code by a simple combinational logic circuit.

As conversion is performed in one clock cycle, this ADC architecture enables very high sampling rates, which are limited only by the settling time of the comparators and the time of response of the associated circuitry. However, their resolution is limited in practice to less than 6 bits (usually 4) due to the exponential growth of circuit complexity and power consumption. Indeed, the number of required comparators is doubled for each extra bit of resolution.

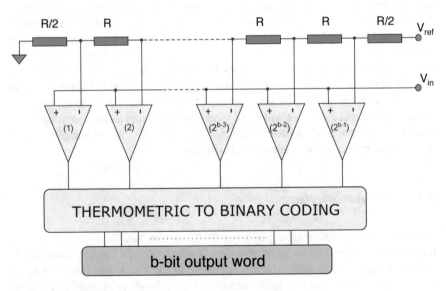

Fig. 5.1 b-Bit flash ADC

5.1.2 SAR: High Resolution, Low Conversion Speed

Figure 5.2 shows the diagram of a successive approximation register (SAR) ADC. This architecture is composed of several subsystems, which area phase control module for timing and clock generation, a sample and hold, a comparator, a shift register, and a digital-to-analog converter (DAC). At the beginning of the conversion cycle, the analog input signal is sampled and maintained at the sample and hold (SH) circuit during the conversion period driven by clock clk_2. Then, the conversion is performed in an inner loop driven by clock clk_1 (which is b times faster). At each cycle of clock clk_1, the input sample is compared to the output of the shift register converted back to the analog domain through the DAC, and a bit is determined (from the most significant bit to the least) by changing the value of the register accordingly. That is, if the comparator is triggered, then the corresponding bit is set to 1, otherwise to 0. After b iterations, the b-bit output word is obtained and a new conversion cycle begins to determine the digital value of the next sample.

5.1.3 Sigma-Delta: Higher Resolution with Low Quantization, Oversampling, and Noise Shaping

Recently, the need for high-resolution analog-to-digital converters (ADCs) with low power consumption, especially for mobile applications, has drawn much attention towards sigma-delta architectures for signal conversion. These ADCs combine low-resolution quantization with oversampling and noise shaping in order to reduce the in-band noise and thus increase the dynamic range. In particular, continuous-time (CT) sigma-delta modulators (SDMs) seem to be an attractive choice because of their inherent anti-aliasing properties and low circuit complexity, among other advantages. In addition, they provide a flexible choice between resolution and bandwidth, which makes them suitable for multi-standard transceiver architectures.

A sigma-delta ($\Sigma\Delta$) ADC is composed of a $\Sigma\Delta$ modulator followed by a low-pass filter and a decimator. The main component of the ADC is the modulator, shown in Fig. 5.3 (for the case of a first order modulator), which is composed of an integrator, a one-bit quantizer, and a DAC in a feedback loop. The order of the modulator is determined by the number of feedback loops with an integrator that are

Fig. 5.2 b-Bit SAR ADC

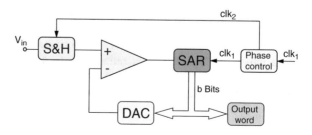

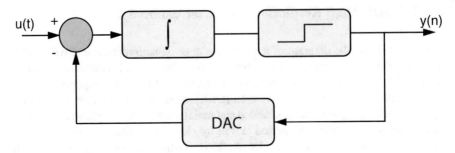

Fig. 5.3 First order $\Sigma\Delta$ modulator

included, which determine the extent of noise shaping that is applied to the input signal. However, first order modulators are generally preferred to avoid stability issues.

Considering the quantizer as an additive noise source, the transfer function for the signal is a *low-pass filter*, while the transfer function for the quantization noise is a *high-pass filter*. Thus, the quantization noise is attenuated at baseband. This noise shaping effect, combined with oversampling, increases the effective signal to quantization noise ratio (SQNR) and therefore the effective ADC resolution. In general, the oversampling ratio used is in the range of 8–16.

5.1.4   Combined Structures

In this section we briefly describe the two main sampling structures used to increase sampling rate and resolution by combining several ADCs in such a way that circuit complexity and consumption are kept as low as possible.

5.1.4.1   Pipelined ADCs: Increased Resolution

A pipelined ADC structure is a combination of K sampling stages connected in a cascaded manner, to increase resolution without sacrificing sampling rate and with a linear increase in complexity and power consumption. Each sampling stage (shown in Fig. 5.4) is composed of an m-bit ADC and DAC, a subtractor, and an amplifier with gain 2^m. The conversion cycle is as follows. The analog input signal is sampled and quantized at the first ADC. Then, this coarse quantization is stored in a register, converted back to the analog domain and subtracted from the input to the ADC. The result, called *residue* is the quantization noise from the sample (and thus $R < LSB$), which is then amplified by a factor $A = 2^m$ to fit the input range of the ADCs and sent to the next sampling stage. This process is repeated in the $K - 1$ following

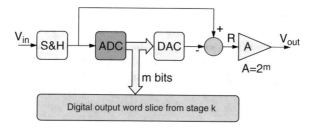

Fig. 5.4 Stage of a pipelined ADC

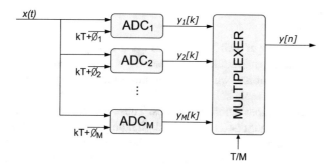

Fig. 5.5 Time-interleaved ADCs

stages, obtaining finer quantization of the input in the corresponding registers, which are then concatenated to obtain a b-bit output word, with $b = Km$.

5.1.4.2 Time-Interleaved ADCs: Increased Sampling Speed

A time-interleaved (TI) ADC is an array composed of several (say M) ADCs working in parallel with time-shifted sampling clocks such that the overall effective sampling rate is increased by M. This increment in the sampling rate is obtained by shifting the relative phase of the clocks that drive each ADC by $\phi = T/m, m = 0, \cdots, M - 1$, where T is the sampling period of each ADC. Then, their output samples are multiplexed at the M times faster rate T/M as shown in Fig. 5.5. These sampling structures allow to increase the effective sampling rate by M with a linear increment in complexity and power consumption.

5.2 Traditional (Narrowband) Compensation Techniques

ADC post-compensation through traditional (narrowband) compensation techniques includes frequently used error correction methods for certain sampling scenarios where simple correction techniques are sufficient to attain acceptable performance. This is the case when sampling narrowband signals at low-to-moderate sampling rates. Under these conditions, error signals are almost static and their behavior can be effectively modeled and corrected at low computational complexity and resources by these methods. In the following sections we will briefly introduce and describe some of them.

5.2.1   Integral Nonlinearity and Differential Nonlinearity Models

The non-ideal behavior of a circuit degrades the overall performance of a sampling system by introducing harmonics and other types of nonlinear distortion as well as increased in-band noise, which reduces the effective number of bits (ENOB) in the converter. A possible solution to reduce this distortion is the use of model-based digital post-compensation techniques. These techniques are generally based on application of another distortion to the digital output of the converter that would cancel out the original distortions present in the device output [3, 4]. They involve two steps: first, the post-compensator is trained using measurement data from the ADC; then, the post-compensator is used at the output of the converter.

Following the definitions of Sect. 2.9.3, and although INL and DNL are usually measured at a single frequency, they are not static in the sense that measurements at different frequencies result in different results. However, their models are popular candidates for narrowband model-based ADC post-compensation. For example, an analysis of non-ideal behavior of an SAR-ADC based on INL/DNL plots is performed in [5], where several sources of errors in this architecture are described and their effect on INL and DNL addressed. Then, this information can be used during the design stage to identify the dominant sources of error by analyzing the INL plots, such that adequate measures to alleviate them can be performed. In [6], INL models particularly suitable for pipelined ADCs are proposed and tested.

Here we present a brief description of the model proposed in [7], where some memory is added to represent frequency dependent changes in INL. As an example, Fig. 5.6 shows the INL error vs. output code for a commercial ADC from Analog Devices [8]. The model proposed in [7] is based on the idea that the INL behavior (as seen in Fig. 5.6) is that of a smooth curve combined with a saw-tooth like function plus noise, such that

$$INL[k] = INL_1[k] + INL_2[k] + INL_3[k], \tag{5.1}$$

where INL_1 can be approximated by an nth order polynomial of the form

$$INL_1[k] = a_0 + a_1 k + a_2 k^2 + \cdots + a_n k^n, \tag{5.2}$$

Fig. 5.6 INL vs. output code for Analog Devices AD9461 ADC at 130 MSPS with a sine wave input of 10.3 MHz

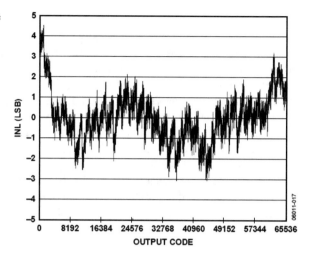

whereas $INL_3[k]$ is white noise. $INL_2[k]$ is assumed to be static and piece-wise linear in $k_p - 1 < k < k_p$ for a limited code interval $p \in [1, 2, \ldots, P]$ and represented by

$$INL_2[k] = b_0[p] + b_1[p](k - k_{p-1}). \tag{5.3}$$

Finally, memory (frequency dependent) effects can be added to the model by replacing the static polynomial representing $INL_1[k]$ by a memory polynomial, which is a parallel form of the Hammerstein box model. Then, an inverse to this model can be identified and applied at the ADC output to cancel distortion.

5.2.2 Look-Up Tables

The idea behind look-up table (LUT) based ADC post-compensation is to use the output samples of the ADC to generate an *index* pointing to certain *memory address* in order to access a particular entry to a *table* where an estimate of the ADC output error (or the "correct" sample value) has been previously stored [9].

This technique involves several procedures that have to be executed before the correction method can be implemented. These include the generation of a reference signal, an estimation method allowing to compute the table entries from the reference signal, and an indexing scheme to determine in which way the address index I is generated from the ADC output samples. Although a certain amount of dynamic effects can theoretically be modeled by LUTs, the exponential increase in the required table size makes practical implementation feasible only when few consecutive samples are used for indexing.

The *indexing scheme* determines in which way the address index I is generated from the ADC output. There are basically three main approaches. The first option is static indexing, where the index I is just the present output sample from the ADC and the correction capabilities of the resulting LUT are able to compensate for static nonlinearities only. State space indexing is a generalization of this method to include information on the input signal dynamics by also adding k past samples to the present output to generate the address index. As a result, some memory can be included in the LUT, but the increment in the required table size grows exponentially as a function of k. The third alternative is phase plane indexing. In this case, the address index is generated by combining the present output sample from the ADC with an estimate of the signal slope, usually determined by a derivative FIR filter. It can also be generalized to use the information of the first k derivatives of the signal. Several techniques have been developed in order to enhance these indexing schemes allowing a reduction in the final LUT size. Some of them can be found in [9] and the references therein.

The *reference signal* used to compute the LUT values can be obtained in diverse manners. These include the use of a (higher quality) reference ADC simultaneously sampling the same signal, generating a signal in the digital domain, and feeding it to the ADC through a high-quality DAC, or estimation of the reference signal from the ADC output through signal processing techniques such as sine-wave fitting algorithms.

Once the reference signal is obtained and the indexing scheme chosen, an *estimation* algorithm can be derived to compute the correction values with which the LUT will be filled.

5.2.3   Dithering

In the field of signal processing, a dither signal is a pseudo-random noise added at the input of an ADC in order to decorrelate the quantization noise from the analog input signal to it [1, 9]. The reason behind this is that although it is usually assumed that the quantization noise is white with uniform distribution, this is not exactly true but an approximation that holds for moderate to high-resolution ADCs. Thus a small amount of dithering makes this assumption more realistic, which can be used in order to avoid correlation of idle patterns with the input signal in low-resolution ADCs such as those used in noise shaping converters like sigma-delta [1]. Several conditions and statistical properties to ensure decorrelation of the noise with the input signal that have been derived under subtractive and non-subtractive dither can be found in [9].

On the other hand, another useful result from the use of dithering is to randomize the DNL and INL patterns of a non-ideal ADC (thus reducing nonlinear distortion). The idea is to add a higher level of white noise (more than an LSB) to the input signal before quantization. As a result, the level of harmonic distortion tones can be decreased at the expense of a higher noise floor [9].

5.3 Novel Compensation Techniques Amenable for Wideband Sampling

Modern wideband communications systems operate today at higher frequencies, with increased bandwidth and dynamic range. This is in part due to the actual trend towards complete digital signal processing systems, even for applications that were partially restricted to the analog domain because of their high operation frequencies and large bandwidth, and has created a demand for analog-to-digital converters (ADCs) of very high speed and low distortion [7, 10]. Under these conditions, nonlinear dynamic behavior and circuit non-ideal effects at the radio frequency (RF) analog front-end (AFE) are strong sources of distortion that severely deteriorate the performance of the system. This is particularly critical at the analog-to-digital converter (ADC), due to propagation of errors in the discrete output of these devices to the rest of the processing chain [11]. In fact, this determines an upper limit in the achievable performance even if the rest of the system is error free. In this sense, efficient solutions for distortion reduction in sampling systems are a key demand.

When sampling wideband signals, the nonlinear dynamics at the ADC produce strong distortion that has in general high order terms and long memory effects, as well as complex dependence between present and past output samples. Thus, the number of parameters of an accurate model to represent it can be dramatically increased, which has a direct impact in the computational complexity of both identification algorithms and post-compensators. Therefore, distortion reduction through modeling and compensation is needed, with the additional challenge of finding an accurate model capable of representing them while keeping the complexity as low as possible. To reduce distortion effects, two tasks must be performed. First, an accurate model of the ADC non-ideal and nonlinear behavior must be proposed and tested. Then, an estimate of the model parameters must be obtained, and only then can this information be used to apply an adequate compensation technique. The parameter estimation phase is therefore critical, i.e., the estimation error must be as low as possible.

An important distinction can be made based on whether the ADC operation must be interrupted or not during the parameter estimation process. Foreground (off-line) calibration relies on an additional training signal known a priori that is used to estimate the compensator parameters [12–14]. This strategy results in more accurate estimates of the parameters involved, but operation must be (periodically) stopped. On the contrary, background (on-line) calibration techniques are capable of directly using the input signal during normal operation of the ADC for estimation and correction of distortion [15]. They can also be divided into blind or semi-blind methods. Blind methods do not require any information on the input signal. Instead, they use either an additional reference ADC or elaborated algorithms exploiting some system knowledge to design suitable cost functions and minimize them adaptively (normally using gradient-based methods) [15, 16]. While the advantage is the resulting great flexibility that can be achieved, their computational complexity can be very high, and they result in lower accuracy.

In the following sections we introduce linearization techniques based on model inversion of Volterra type models, as well as estimation and correction of mismatch errors in TI-ADCs based on the results from [13, 14, 17, 18], and [19].

5.3.1   Model Inversion: Nonlinear Dynamic Models

It is well known that systems presenting weak nonlinearities usually have a Volterra representation. Moreover, discrete-time systems with fading memory can be approximated arbitrarily well by a discrete-time Volterra model (DTVM), if adequate orders are chosen [20–22]. In addition, a Volterra model can be p-linearized by a model of similar complexity [20]. Based on this knowledge, we introduce a Volterra-based modeling and post-compensation technique for the case of CT-SDMs that, as we will show later on, is also a very good fit for high-performance commercial ADCs. We also present a behavioral model of the modulator that provides extra information allowing to select and design an accurate DTVM compensator with reduced complexity.

As previously mentioned, sigma-delta ADCs combine low-resolution quantization with oversampling and noise shaping in order to reduce the in-band noise and thus increase the dynamic range. In particular, CT SDMs seem to be an attractive choice because of their inherent anti-aliasing properties and low circuit complexity, among other advantages [1]. Sigma-delta structures have been proposed for many applications, including DVB-T (Digital Video Broadcasting-Terrestrial) [23–25] and Bluetooth [26]. In addition, they provide a flexible choice between resolution and bandwidth, which makes them suitable for multi-standard transceiver architectures combining for example GSM/WLAN/bluetooth [27, 28].

5.3.1.1   Behavioral Model of Continuous-Time $\Sigma\Delta$ Modulators

In order to obtain an adequate structure for the compensator it is first necessary to understand the non-ideal behavior of CT SDMs. Several partial studies have been performed in the literature. For example, the issue of the nonlinearity in the integrator is considered in [29–31] modeling the quantizer effects as an additive noise source. In [30] and [31] a Volterra model is developed following the additive noise assumption for SDMs with multibit quantization. In other line, in [32] the non-ideal effects on the DAC are addressed and in [33] a new interpretation on the quantizer effects is discussed. However, none of the mentioned works offers a complete description of the SDM including all these effects jointly.

Therefore, we propose efficient finite Volterra based post-compensation schemes for this type of converters in order to maintain a low complexity for the necessary on-line processing. In order to achieve this, we first develop a behavioral model for such devices by studying the different elements composing the modulator and considering all their effects on the system. Since the behavioral model represents in general weak nonlinearities, it is possible to consider a finite Volterra model to capture all the mentioned characteristics. This Volterra model allows to design and use a post-compensation model and the corresponding parameter estimation techniques. In the following, we consider a model for each block introducing the real effects that preclude SDM ideal behavior, with the aim of finding an equivalent block oriented model that represents the non-ideal SDM.

Integrator nonlinearities is one of the main factors that limit the maximum achievable SNR in a sigma-delta analog-to-digital converter (SDC). The linearity of the integrator is limited by the effect of the nonlinear trans-conductance of the operational amplifier which appears in the output integrator current [30]. From this point of view, continuous-time integrators can be approximated as a cascade of a static nonlinear operator and an ideal integrator device [29]. In the case of fully differential architectures, which are usually used for amplifiers to implement the integrator, the even terms can be neglected due to the common mode rejection ratio. Assuming this structure here, the nonlinear operator will contain only odd terms [30, 31]. In general, it can be assumed that a third order nonlinearity is sufficient to capture the main nonlinear effects [30, 31]. Thus, we can assume that the static nonlinear operator is given by

$$f(x) = ax - bx^3, \tag{5.4}$$

where x is the input signal to the integrator, and a and b are constants depending on the trans-conductance of the transistors used in the integrator. The derivation of (5.4), for different integrator architectures, can be found in [31].

Non-ideal behavior of the feedback DAC has to be taken into account when a precise model is required. In order to analyze these effects, we present a model based on the results presented in [32]. Given a differential binary signal $y(n)$ at the input of a single-bit DAC, we can define its output as

$$y(t) = \sum_{n=-\infty}^{\infty} y_0(n)h(t - nT), \tag{5.5}$$

where T is the clock period, $y_0(n)$ is defined as:

$$y_0(n) = \begin{cases} +V_{ref} & \text{if } y(n) \geq 0 \\ -V_{ref} & \text{if } y(n) < 0 \end{cases}$$

with V_{ref} as the reference voltage for the DAC, and

$$h(t) = \begin{cases} 1 & \text{if } 0 < t < T \\ 0 & \text{otherwise} \end{cases} \tag{5.6}$$

is an ideal sample and hold device for the DAC. This model can be modified to include the additional effects that exist in real circuits, such as propagation delays and non-instantaneous swing between positive and negative reference voltage. To that purpose, the function $h(t)$ in (5.6) can be replaced by a first-order filter response, generating an exponential voltage settling at the output of the DAC, i.e.,

$$h(t) = 1 - e^{-t/\tau} \qquad (5.7)$$

which has the following Laplace transform:

$$H(s) = \frac{1}{1 + s\tau}, \qquad (5.8)$$

where τ is the time constant of the circuit, determined by the slew rate of the transistors that limits the time response of the DAC. Note that the actual response of the DAC is still linear.

Smooth nonlinearity of the quantizer is the last requirement for development of a SDM Volterra model. In a single-bit first-order CT SDM, the one-bit quantizer (usually a comparator) can be modeled using the sign function, which is strongly nonlinear. However, the noise shaping of the quantization noise and the feedback nature of the SDM linearize the quantizer in the signal bandwidth [33]. In fact, little quantization noise occurs in the signal band, and out-of-band quantization noise is filtered out by the decimation filter. The nonlinearity of a static transfer function can be drastically reduced using an additive dither signal[1] at the input of the nonlinear element [33]. For this purpose, the dithering signal must satisfy the condition of having a fundamental frequency much higher than the input signal bandwidth. If the dither signal is a sinusoid of amplitude A_d, and its frequency is much higher than the input signal bandwidth, then the output signal (after a low-pass filter) can be written as [33]:

$$y_{lp} = \frac{2}{\pi} sin^{-1} \left(\frac{x}{A_d} \right) \approx \frac{2}{\pi A_d} x; \qquad x << A_d. \qquad (5.9)$$

Therefore, the dither signal linearizes the quantizer in the signal bandwidth. In order to generate a dither signal, an oscillator consisting of a high dynamic gain followed by a static nonlinear element is needed. That is the case of the integrator followed

---

[1] Unlike the concept of a dither signal discussed earlier as a pseudo-random noise added at the input of an ADC in order to de-correlate the quantization noise from the analog input signal to it, the same technique is used in control of nonlinear systems but with a different objective. In this case, a high frequency sinusoidal signal is used to change the behavior of a nonlinearity in such a way that an averaging effect takes place. This is due to the convolution between the nonlinearity and the amplitude distribution of the sinusoidal signal. It can be shown that the nonlinear element, usually a strong or discontinuous nonlinearity, behaves as a smoother nonlinear element in the lower frequency range. Here, we use the second interpretation of dithering.

by a comparator in a single-bit CT SDM, combined with the oversampling and feedback. Hence, in an SDM, the quantizer input contains a dithering signal. As a general conclusion, the quantizer model can be well described by a weakly static nonlinear system. Hence, if $x(t)$ is the input to the comparator, we can model its output with an pth order polynomial

$$b[x(t)] = x(t) + k_1 x^2(t) + \cdots + k_{p-1} x^p(t), \tag{5.10}$$

where the k_i ($i = 1, \ldots, p - 1$) are the polynomial coefficients. As suggested by (5.9), the output of the comparator is almost linear, and then $p \leq 3$ should suffice.

In the case of a SDM with a multibit quantizer, the usual assumption is that the multibit analog-to-digital converter (ADC) in the forward path can be replaced by an additive noise source [1, 30, 31]. This noise source is assumed to be white with uniform distribution on the interval $-LSB/2$ to $+LSB/2$, where LSB is the quantization step. However, this is still an approximation for high-resolution ADCs. If we consider a low-resolution multibit quantizer, i.e., up to 4 bits (which is usually the case in SDMs), then another approach should be considered. Let us consider an SDM with a multibit quantizer. Hence, assuming a flash[2] B-bit quantizer with thermometric coding, the output of the multibit DAC at the feedback branch can be written as:

$$y(t) = b_1(t) + b_2(t) + \cdots + b_B(t), \tag{5.11}$$

where b_i is the i-th bit corresponding to the output of comparator i, with different reference voltages for each $i = 1, 2, \ldots, B$. Since a single bit quantizer in an SDM is linearized in the band of interest (and therefore it can be modeled with a weakly static nonlinearity), each bit b_i in (5.11) can be represented arbitrarily well with a polynomial such as the one represented in (5.10). Then, $y(t)$ can be represented as a sum of polynomials, which in turn is a polynomial as well. Therefore, the multibit quantizer can also be modeled as a weakly static nonlinear system.

The complete behavioral model is obtained as follows: first, the integrator in Fig. 5.3 can be replaced by a third-order polynomial $P(\cdot)$ followed by an ideal integrator (i.e., a linear filter $I(s)$). Then, the DAC in the feedback loop can be represented by a filtered version of the output signal $y(t)$ through a linear filter $H(s)$. Finally, the quantizer can be considered almost linear in the signal bandwidth, and so we can model it with a weak static nonlinearity $N(\cdot)$. As a result, we get the equivalent functional diagram shown in Fig. 5.7.

---

[2]For example, a flash analog-to-digital converter is a natural choice because of the high sampling rate of the system and the low resolution required [34].

Fig. 5.7 Equivalent
functional diagram of a SDM

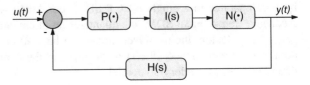

5.3.1.2   Volterra Model and Post-compensation of a Continuous-Time ΣΔ Modulators

In general, the output of a DTVM, at instant k, can be expressed as [21]:

$$y(k) = \Phi(y(k-1), \ldots, y(k-p), u(k-1), \ldots, u(k-q)), \qquad (5.12)$$

where the choice of the function $\Phi(\cdot)$ and the parameters p and q define the model.

In the case of the system depicted in Fig. 5.7, an equivalent Volterra representation can be formulated as follows. First, we consider that the CT SDM can be approximated arbitrarily well by a discrete-time system with a sufficiently high sampling rate. Then, we assume, without loss of generality, that $I(s)$ and $H(s)$ are linear systems with finite memory (as is the case for most physical systems). Thus, we consider a length M_I for $I(s)$ and a memory of M_H for $H(s)$. Also, it is possible to assume that signal components that have already passed through the feedback path will not re-enter the loop.[3] This assumption is based on two facts. On one hand, the nonlinearity $P(\cdot)$ will mix up some of these components to higher frequencies out of the band of interest. On the other hand, the remaining components would be so far away in the past that their effect can be considered negligible on the present output $y(n)$. Therefore, the system will also have a finite memory M such that $M \cong M_I + M_H$, so fading memory can be assumed. On the other hand, the nonlinear effects of the system are frequency dependent and vanish when the excitation is removed.

Since the system presents finite memory, we choose $p = 0$ in (5.12), i.e., nonlinear FIR (NFIR) systems and focusing on the family of analytic continuous functions $\Phi(\cdot)$ (which can be expanded into Taylor series), it is possible to define DTVMs analogous to the continuous-time Volterra models. In such case, the integrals are replaced by discrete convolution sums and the system response becomes

$$y(k) = \sum_{l=1}^{L} y_l(k), \qquad (5.13)$$

---

[3] A similar approach was used in [33] to model the dynamic nonlinearities in a radio frequency power amplifier (RF PA).

where the first term, given by

$$y_1(k) = \sum_{i_1=0}^{\infty} \alpha_1(i_1)u(k - i_1) \tag{5.14}$$

corresponds to the linear convolution model, and the higher order terms can be written as:

$$y_l(k) = \sum_{i_1=0}^{\infty} \cdots \sum_{i_l=0}^{\infty} \alpha_l(i_1, \ldots, i_l)u(k - i_1) \cdots u(k - i_l). \tag{5.15}$$

Furthermore, any nonlinear function can be approximated by a polynomial of sufficiently high order leading to finite DTVMs, which are composed of a moving average linear model of order M and a polynomial nonlinearity of degree N. For the SISO case, the input–output relationship of such systems is given by

$$y(k) = y_0 + \sum_{n=1}^{N} v_M^n(k), \tag{5.16}$$

where

$$v_M^n(k) = \sum_{i_1=0}^{M} \cdots \sum_{i_n=0}^{M} \alpha_n(i_1, \ldots, i_n)u(k - i_1) \cdots u(k - i_n). \tag{5.17}$$

At this point, we can extract further information about the Volterra representation of the SDM structure in Fig. 5.7. As previously mentioned, the linear filter in the feedback path generates cross terms of the input signal to the output. Also, the feedback system with Hammerstein–Wiener model in the forward path suggests that a more general model than Hammerstein–Wiener is needed to represent the SDM. Finally, there are two different linear filters, i.e., $I(s)$ and $H(s)$, so a model with at least two linear dynamics is required.

As a first step, we analyze the performance of a memory polynomial model (MP) for the compensator (shown in Fig. 5.8). This model is a generalization of a Hammerstein system [35], where multiple linear filters are allowed after the static nonlinearity, and has been commonly used for post-compensation in several systems (including the INL models in [7] discussed earlier in this chapter). Assuming an Nth order polynomial and an FIR filter of length M, the compensator output is given by

$$\hat{y}^{MP}(k) = \sum_{n=1}^{N} \left(\sum_{m=0}^{M-1} \alpha_{nm} y^n(k - m) \right), \tag{5.18}$$

Fig. 5.8 Block diagram of an
MP model

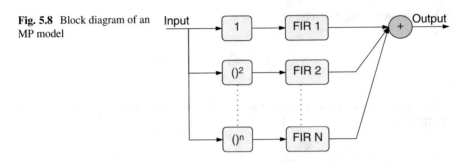

where $y(k)$ is the output of the SDC, $\hat{y}^{MP}$ is the output of the compensator, and α_{nm} are the parameters involved. This structure has the advantage that the output signal is linear in the unknown parameters (i.e., the terms α_{nm}), which renders efficient parameter estimation possible through least-squares methods [36].

As an alternative to the MP model, we introduce a modified generalized memory polynomial (MGMP). This model is the transposed of the block diagram shown in Fig. 5.8, with the power terms after the FIR filter in each parallel branch. It can be shown that in this case, the input–output relationship of the model is

$$\hat{y}^{MGMP}(k) = \sum_{n=1}^{N} \left(\sum_{m=0}^{M-1} \alpha_{nm} y(k-m) \right)^{n} . \tag{5.19}$$

Note that this model includes cross terms among the samples $y(k-m)$ at its output. As will be shown later on, this model has not only better representation capabilities but also enhanced generalization properties, i.e., it produces better compensation results during the validation phase where the estimated model is tested for input signals which have not been used for training.

5.3.1.3   Performance in Post-compensation

The proposed post-compensation method was calibrated using signals obtained from a transistor level circuit model by transient simulations in Spice [18]. This circuit model was also compared to the behavioral model showing good agreement between both, and then used to generate input–output data for the design of both post-compensators and the estimation of the parameters involved. This provides realistic simulation data, leading to general and reliable results when evaluating compensation performance. Our device under test (DUT) has a sampling frequency of 100 MS/s, over a bandwidth from DC to 1 MHz, determining an over-sampling ratio OSR > 50 and a resolution of over 7 bits. Return-to-zero (RZ) coding is used in the feedback loop, which is known to reduce errors in the modulation [1]. A latch outside the loop codes the signal in nonreturn-to-zero (NRZ) format. The architecture is fully differential and the design uses 180 nm CMOS technology with the transistor model provided by manufacturer MOSIS. The performance of the compensation method was simulated in MATLAB for different sets of input–output signals provided by the circuit model of the SDM. An ideal SDM was simulated in MATLAB both to estimate the parameters of the compensators and to measure their performance through the generation of the ideal SDM output signal y_I. The input signal used to excite this ideal modulator is imported to MATLAB from the circuit simulator software, so the output of both models can be compared. In our simulation studies, different sinusoidal single-tone and multitone input signals are used to excite the circuit, as usually found in the literature [7, 9, 37].

The tests using single-tone signals show that a large improvement can be obtained by compensation using both models, but it is not clear which one offers a better representation for the dynamics of the system [18]. For this purpose, a more general signal has to be used as excitation for the circuit model of the SDM. Therefore, we chose a multitone signal (MT) with four tones at 100, 200, 400, and 800 kHz. This signal does not only cover most of the signal bandwidth but frequencies are chosen in such a way that harmonics due to different frequency components do not overlap. Figure 5.9 shows the spectrum of the input signal and the output of the DUT before and after compensation using both models. We see that both models achieve a good cancellation of harmonic distortion. However, compensation using the MGMP model achieves better cancellation of harmonic distortion, including the harmonic peak at 600 kHz in the spectrum that is only successfully cancelled by this model. As discussed earlier, this is a result of its improved representation capabilities.

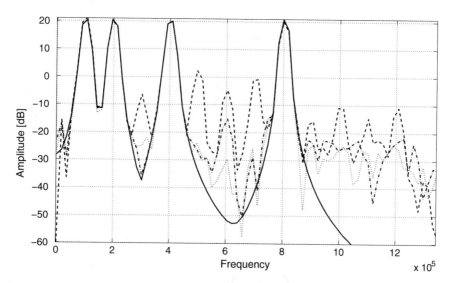

Fig. 5.9 Spectrum of the input signal (full line) and the output of the DUT before (dashed line) and after compensation (MP: dot-dash, GMP: dots)

5.3.2   *Wideband Compensation of High-Performance ADCs*

In order to validate the practical feasibility of the proposed post-compensator and verify its enhanced modeling and generalization properties, we include results from applying it to reduce nonlinear distortion effects in a high-performance commercial ADC using input–output data from actual measurements. The DUT is the AD9461 ADC from Analog Devices [38].

The AD9461 is an ADC with a resolution of 16 bits and a maximum sampling frequency of 130 Msps. This ADC was chosen due to its high-performance characteristics, which combine high conversion speed and resolution over a wide range of frequencies. The converter data sheet [38] specifies a maximum integral nonlinearity INL $= \pm 5$ LSB. According to (2.54), the reduction in effective resolution due to INL is 3.1240 bits, which implies a typical effective resolution of 12.876 bits. Moreover, according to the manufacturer, the maximum measured SINAD (equal to 78 dB) occurs for a sinusoidal signal of 14.5 dBm at 10 MHz, which implies an ENOB of 13 bits, in the best case. However, many different error effects contribute to overall ENOB, such as INL/DNL, eventual clock impurities, dynamic nonlinearity, etc. In addition, a lower reference voltage in the ADC core also deteriorate the converter performance [38] and thus an even lower ENOB should be expected. This information suggests that the performance of this converter can be considerably improved in terms of the effective resolution, enhancing the performance achieved by the circuit design via post-compensation.

Figure 5.10 shows the measurement set-up composed of the ADC card connected to a PC through a data acquisition card included in the test kit. The left card includes the converter, the signal conditioning circuits, and the clock circuitry. The right card is a FIFO data acquisition card with a memory of 32 kB, synchronism and temporization circuits, and USB interface for a PC link.

Fig. 5.10 ADC data acquisition kit

The best possible scenario for compensation of an ADC in terms of resolution enhancement is when sampling a sinusoid signal of known frequency. In this case, either the compensator model fails to capture the dynamics of the system and thus does not linearize the output of the ADC, or it does capture the dynamic behavior of the DUT and achieves an increment in the ENOB. In the latter case, the improvement is most likely to be higher than that obtained for a sampled signal of frequency different than the signal used for training. Hence, as the first step in the process of post-compensation for the DUT, we test the models presented in Sect. 5.3.1.2 for a sampled signal of 36 MHz. This signal was low-pass filtered with a sixth-order passive Butterworth filter in order to enhance its spectral purity beyond that of the analog signal generator, and represents the worst case among the measurements taken in terms of SINAD. The SINAD at the output of the filter (measured with a spectrum analyzer) was 72.5 dBfs, and the SINAD at the output of the DUT (measured with the ADSim software from Analog Devices) was 64.5 dBfs. Therefore, an increment of 8 dB is possible, equivalent to 1.4 bits in the ENOB. This increment is close to 2 bits for the rest of measurements available at other frequencies. From the 32,000 input–output data points obtained, the first 20,000 were used to train both a MP and MGMP compensator and the remaining 12000 were used for validation.

First, several compensators of each type were trained for a polynomial of order 3, as a function of the length of the FIR filters M. The results are shown in Fig. 5.11. Similar experiments can be conducted by varying the polynomial order, such that all model structure parameters (i.e., polynomial order and memory length) can be obtained. As can be seen, the MGMP compensator outperforms the MP. This was likely to be expected, as the MGMP model is more general in the sense that cross terms between past samples are taken into account and it therefore has better

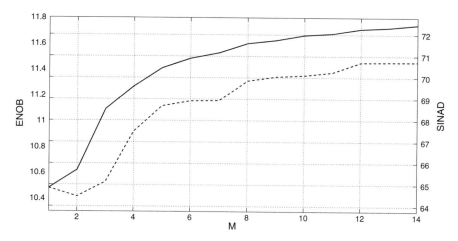

Fig. 5.11 ENOB and SINAD vs. M for MGMP compensator (solid line) and MP (dashed line) using third-order polynomials

representation capabilities. Hence, the following analysis focuses on the MGMP compensator.

The general idea behind the proposed compensation strategy is to train a compensator using as much information as possible about the ADC dynamics in all operating regions (i.e., over a wide range of frequencies). In this manner, the post-compensator should be able to improve the performance in terms of resolution enhancement independently of the frequency of the sampled signal. In order to do so, we propose the use of a more complete signal for training, obtained by concatenation of input–output data from filtered sinusoid signals of different frequencies. In this case, it is important to select a number of tones greater than the memory of the system. Therefore, we use 17 sinusoids for training the compensator, and we validate the post-compensator technique for 8 different frequencies to those used for training.

Figure 5.12 shows the SINAD improvement as a function of frequency. It can be seen that the compensator is able to enhance the SINAD for all the measured frequencies. Figure 5.13 shows the spurious free dynamic range (SFDR) of the DUT and that of the compensator output, where an average improvement of 10 dB is obtained for all cases. Figure 5.14 shows the output spectrum of the DUT before and after compensation for an input sinusoid of 23 MHz, not included in the training sequence. It can be seen in the figure that distortion is successfully reduced, particularly in terms of SFDR enhancement. As stated in [39], MGMP provides the best compensation performance among the available discrete Volterra models for ADC distortion reduction. In addition, performance could be further improved by allowing different memory lengths for the filters in the model corresponding to the different power terms of the polynomial. In this manner, it is possible to optimize the

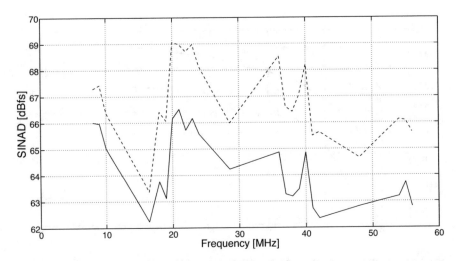

Fig. 5.12 SINAD vs. frequency of the DUT (solid line) and the compensator output (dashed line)

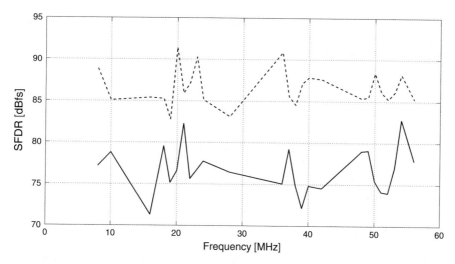

Fig. 5.13 SFDR vs. frequency of the DUT without compensation (solid line), and after compensation (dashed line)

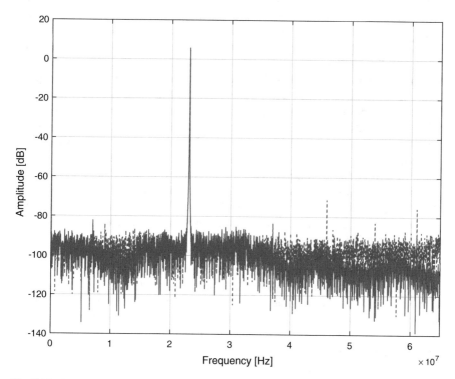

Fig. 5.14 Output spectrum of the DUT without compensation (dashed line), and after compensation (solid line)

number of coefficients in the model such that the estimation accuracy is enhanced
and the complexity reduced. In that case, it is shown that more than 2 bits can be
recovered in the ENOB.

5.3.3 Mismatch Errors in TIADCs

As a consequence of its wide spectrum, modern communication systems can have
bandwidths of up to several GHz. There exists therefore an ever increasing demand
for high-performance and high-speed analog-to-digital converters (ADCs) in order
to comply with the sampling requirements of modern wideband communications
systems and standards like OFDM, UWB, LTE, 5G, optic transceivers, and cognitive
radio [2, 40–42]. However, there is also traditionally a trade-off between the
achievable resolution and sampling speed [1, 2]. Among the alternatives, time-
interleaved (TI) sampling ADCs are a promising solution that has become a trend
and active research topic, as they are key to sample the signals at the required rates.

As previously stated earlier in this chapter, a TI-ADC is an array composed
of several (say M) ADCs working in parallel with time-shifted sampling clocks
such that the overall effective sampling rate is increased by M. However, due to
inaccuracies inherent to the manufacturing process that prevent the component
ADCs from being exactly equal, there are specific mismatches that can severely
deteriorate the performance of the whole system. Thus, addressing three typical
mismatches: gain, offset and timing skew, estimation and correction is required
[1, 2, 11, 43]. Gain mismatch errors occur when the amplitude ratio between analog
input and digital output differs for each ADC, whereas offset mismatch is due to
different DC values at the output of each ADC, even when the input is set to
zero. Finally, timing mismatch errors cause the output signal to be periodically
but nonuniformly sampled [13, 44]. Unlike offset and gain mismatches, which are
static, distortion due to timing mismatch is dynamic (i.e., signal dependent) and
requires additional signal processing with higher computational complexity. While
offset and gain mismatches are quite straightforward to estimate and cancel [13, 45],
background on-line timing mismatch estimation remains a challenge and motivates
active research [12, 15, 16, 41, 46–48].

Several alternatives have been recently proposed to tackle the problem [46, 47].
For instance, signal processing can be performed in the digital domain, or in both
analog and digital domains (mixed mode solutions) [12, 41]. In [41], the mean
squared error (MSE) in detection is measured in the digital domain, while delays on
the clock paths to each ADC are adjusted in the analog domain through adjustable
delay paths until the MSE is minimized. The drawback of solutions involving
analog processing (either for estimation or correction) is that they require additional
hardware. In addition, their estimation and correction capabilities are bound to the
precision of this extra hardware. On the other hand, fully digital techniques only
require additional digital processing and can thus be adapted to any ADC [12].

For instance, in [16], the distortion part of the ADC output signal is blindly estimated by doing several Hilbert transforms, frequency shifting, and folding operations and uses an LMS algorithm for each channel ADC to minimize a cost function. In [48], estimation of timing mismatch is obtained through derivative filters, where an LMS algorithm using Taylor approximations is used for the cost function. In this case, the method is tested with a sampling frequency of 3.6 GHz and timing mismatch in the order of a few picoseconds, which implies that the method is accurate when the timing mismatch is low, in accordance with the results in [13], where it is shown that linear interpolation and spline interpolation are sufficient in this case. However, for larger timing mismatch, higher order interpolation techniques are needed. Finally, as an alternative, background on-line (adaptive or not) estimation and correction can be obtained by means of pilot-based methods [19].

Pilot-based on-line calibration methods gather the advantages of both blind and foreground techniques. Pilot signals are transmitted signals known at the receiver which are used in systems and standards such as OFDM, IR-UWB, etc., for different tasks such as channel estimation, synchronization, or carrier frequency offset estimation [42, 49]. The idea behind this approach is to use these pilot tones or symbols along with any particular knowledge on the system itself to get on-line accurate estimates of the distortion parameters (mismatch errors in the case of TI-ADCs). This enables lower computational complexity and higher accuracy when compared to blind methods, as well as tracking capabilities to changes in the parameters while maintaining normal operation of the sampling stage as opposed to foreground calibration.

After the mismatch errors have been accurately estimated, this estimates can be used to effectively cancel their effects. In the following subsections we introduce possible estimation and correction techniques.

5.3.3.1 Signal Representation

Let $r(t)$ be the received signal and T_c the overall high-speed sampling period, $T = MT_c$ the sampling period at each channel ADC, and $\tau_m^I = mT_c$ the ideal sampling shift for the mth ADC. Then, the ideally sampled signal without mismatch errors is

$$r_m^I(kT) = r^I(mT_c + kMT_c) = r_m^I(kMT_c) = r^I(nT_c) \tag{5.20}$$

with $m = 0, 1, \ldots, M - 1$ and $k, n \in Z$, where $r_m^I(kMT_c)$ is the polyphase decomposition of the signal [50], and $nT_c = mT_c + kT$. If we add the gain, offset and timing mismatch errors G_m, O_m and Δt_m, we get

$$r_m(kT) = G_m(kT)r(mT_c + kMT_c + \Delta t_m) + O_m(kT)$$
$$= G_m(kT)[r_m^I(kT) + \Delta r_m(kT)] + O_m(kT). \tag{5.21}$$

If we define

$$G_m(kT) = G(mT_c + kT) = G(nT_c)$$

$$O_m(kT) = O(mT_c + kT) = O(nT_c). \tag{5.22}$$

Then

$$r_m(kT) = G_m(kT)[r_m^I(kT) + \Delta r_m(kT)] + O_m(kT) \tag{5.23}$$

and considering that $nT_c = mT_c + kT$, we can write the output of the TI ADC as:

$$r(nT_c) = G(nT_c)[r^I(nT_c) + \Delta r(nT_c)] + O_m(nT_c). \tag{5.24}$$

From here on, we use the notation $r_m[k]$ for $r_m(kT)$ and $r[n]$ for $r(nT_c)$.

5.3.3.2 Off-Line Estimation

In this section we propose an estimation method which is simple and capable of estimating the parameters accurately even for large mismatch errors. We assume a sinusoid training sequence to illustrate the method (i.e., $x(t) = \sin(2\pi f t + \phi)$, $f <$ $2M/T$, such that at least two samples per period are taken in each channel in order to avoid aliasing in the training phase). The estimation of mismatch errors is performed off-line, and thus the exact training signal can be estimated as well without much effort. For example, in [14], the training sequence was estimated off-line from a set of measurements of the ADC output by minimizing the RMS error between the acquired data and an ideal ADC.

Consider the output of the mth ADC in the interleaved array with gain, offset, and timing mismatch errors. If we compare it with the ideal sample $r_m^I[k]$, we get

$$E_{O_m}[k] = r_m[k] - r_m^I[k] = (G_m - 1)\, r_m^I[k] + G_m \Delta r_m[k] + O_m. \tag{5.25}$$

As demonstrated in [13], the LS estimate of the offset for the mth ADC is given by

$$\hat{O}_m = \frac{1}{L} \sum_{k=1}^{L} E_{O_m}[k] = \frac{1}{L} \sum_{k=1}^{L} \left[(G_m - 1)\, r_m^I[k] + G_m \Delta r_m[k] + O_m \right]. \tag{5.26}$$

It can be shown that if $r_m^I[k]$ has zero mean, so does $\Delta r_m[k]$. Then, the estimator is unbiased and therefore, we have from (5.26) that if L is large enough and $r_m^I[k]$ has zero mean, then $\hat{O}_m \cong O_m$. Note that the offset estimator becomes unbiased for high values of L, and it therefore is asymptotically unbiased.

Now consider the gain error function defined as

$$
E_{G_m}[k] = \frac{r_m[k] - \hat{O}_m}{r_m^I[k]} = G_m + \frac{G_m \Delta r_m[k] + O_m - \hat{O}_m}{r_m^I[k]}. \tag{5.27}
$$

Similarly, we can take L data samples and formulate the least squares estimate for the gain mismatch error G_m as:

$$
\hat{G}_m = \frac{1}{L} \sum_{k=1}^{L} E_{G_m}[k] = G_m + \frac{1}{L} \sum_{k=1}^{L} \frac{G_m \Delta r_m[k] + O_m - \hat{O}_m}{r_m^I[k]}. \tag{5.28}
$$

Having unbiased estimates of O_m, we see from (5.28) that if L is large enough, provided $r_m^I[k] \neq 0$, then $\hat{G}_m \cong G_m$. If $r_m^I[k] = 0$, then that sample is simply discarded for the estimation of G_m.

Let us now define the remaining error due to timing mismatch after correcting for offset and gain mismatches

$$
E_{\Delta t_m}[k] = \frac{1}{\hat{G}_m} \left[r_m[k] - \hat{O}_m \right] - r_m^I[k] \cong \Delta r_m[k]. \tag{5.29}
$$

Assuming a sinusoid input signal, i.e., $x(t) = \sin(2\pi f t + \phi)$, we can consider a simple and efficient manner to estimate the timing mismatch as follows.

According to [1], the maximum amplitude error ΔA due to a shift Δt in the sampling instant occurs for the higher frequency component of the input signal. As we are using a sinusoid training signal, we should chose a frequency close to half the channel sampling frequency, and the maximum error will occur near the zero crossings, where the slope of a sinusoid signal is maximum. It can therefore be computed from a vector of L data samples as

$$
\hat{\Delta t}_m = \frac{\max\left[E_{\Delta t_m} \right]}{2\pi f_0}, \tag{5.30}
$$

where f_0 is the frequency of the input signal. The derivation of (5.30) can be found in [1].

5.3.3.3 On-Line Estimation

According to [45], offset and gain errors can be adaptively estimated on-line as the mean and variance of the individual ADC outputs, respectively.

Then, considering a balanced source of information (i.e., the amount of transmitted ones and zeros is roughly equal), the input signal to the ADC has zero mean and the *offset* mismatch can be calculated by directly averaging K samples at the output of each channel ADC

$$\hat{O}_m = \frac{1}{K-1} \sum_{k=1}^{K} r_m[k]. \tag{5.31}$$

As for the *gain* mismatch, it can be computed as the variance of each ADC output

$$\hat{G}_m = \frac{1}{K-1} \sum_{k=1}^{K} (r_m[k] - \mu)^2, \tag{5.32}$$

where $\mu = 0$ if we consider the offset has been previously cancelled. In that case, (5.32) reduces to

$$\hat{G}_m = \frac{1}{K-1} \sum_{k=1}^{K} (r_m[k])^2 \tag{5.33}$$

otherwise, μ in Eq. (5.32) can be computed using Eq. (5.31). Note that both (5.31) and (5.32) can be adaptively updated (on-line) by adding the next sample $r_m[k+1]$ to the calculations. This enables not only on-line background estimation but also tracking capabilities to changes in the parameters that could arise due to temperature variations, aging, etc.

5.3.3.4   Compensation

Compensation of gain and offset mismatch errors in TI ADCs can be easily accomplished [13] by first subtracting the estimated offset from the output samples of each ADC, and then multiplying by the inverse of the estimated gain error.

As the timing shifts have previously been estimated and the overall ideal sampling frequency is known, we have available the following data,

- The actual instants where samples are being taken in each ADC ($t_m[k] = kT + \Delta t_m$).
- The value of the samples taken at the actual time instants after correcting for offset and gain error $\hat{r}_{adm}[k]$.
- The ideal sampling instants with no timing mismatches $t[n] = nT/M$.

Therefore, we can estimate the ideal samples that would have been taken at the ideal time instants through interpolation between the available samples using the available timing information. To that purpose, we propose the use of a simplified form of the Lagrange interpolation polynomial, as discussed in [13]. This leads to a time-varying filter implementation. The coefficients of the resulting time-variant filter are known, so the computational complexity is in the order of that of an FIR filter. However, other compensation methods such as fractional delay filters [51] or spline interpolation [52] could also be used.

Lagrange interpolation is composed of a set of N orthonormal polynomials of Nth order. Then, given a function $f(t)$ and a data set $f(t_1), \ldots, f(t_N)$, the nth polynomial described by

$$P_n(t) = \prod_{\substack{j \neq n \\ j=1}}^{N} \left[\frac{t - t_j}{t_n - t_j} \right] \tag{5.34}$$

satisfies $P_n(t) = 1$ if $t = t_n$ and $P_n(t) = 0$ for $t = t_j, j \neq n$.

Hence, the polynomial defined as

$$P(t) = \sum_{n=1}^{N} [P_n(t) f(t_n)] \tag{5.35}$$

is an approximation for $f(t)$ in the given interval, defined by the available data set.

It is clear from (5.34) and (5.35) that $P(t_n) = f(t_n), n = 1, \ldots, N$. The approximation of $f(t)$ by $P(t)$ will be better at the center of the interval of data samples used for fitting the polynomial (around $t_{N/2}$) and the approximation will be more accurate for higher N. However, if N is high, convergence problems will arise in the borders, i.e., around t_1 and t_N. In TI ADCs, $f(t_n)$ is the output sample of the TI ADC array after correcting for gain and offset mismatches, and $t_n = nT + \Delta t_m$ is the sampling instant at the periodic but nonuniform time interval defined by the sampling frequency of each ADC and the (previously estimated) sampling shift Δt_m in the mth ADC. Hence, we use the Lagrange interpolation polynomial to estimate $f(t)$ at the correct sampling instant t. From (5.34) and (5.35) that procedure requires N^2 multiplications, $N(N-1)$ divisions, and N sums for each sample. Thus, the complexity of such approach can be quite high for large N. Since we know a priori the values of subsequent t_n, thanks to the estimation of Δt_m for each ADC, a simpler implementation can be obtained after some considerations inherent to TI sampling. Given the periodic nature of the sampling time instant deviations in TI ADCs, the following (more detailed) notation can be used instead

$$t_n = nT_s + \Delta t_n \tag{5.36}$$

$$t_j = jT_s + \Delta t_j.$$

Note that for M TI ADCs, the time deviation will be the same every M samples, and therefore we can write

$$t_{n+M} = (n + M)T_s + \Delta t_n \tag{5.37}$$

$$t_{j+M} = (j + M)T_s + \Delta t_j.$$

Therefore, if we estimate the sample at the ideal time instant t at the center of the time window

$$t = (N/2)T_s \tag{5.38}$$

$$t + MT_s = (N/2 + M)T_s.$$

Then,

$$
\begin{aligned}
P_n(t + MT_s) &= \prod_{\substack{j \neq n \\ j=1}}^{N} \left[\frac{t + MT_s - t_{j+M}}{t_{n+M} - t_{j+M}} \right] \\
&= \prod_{\substack{j \neq n \\ j=1}}^{N} \left[\frac{t + MT_s - (j + M)T_s - \Delta t_j}{(n + M)T_s + \Delta t_n - (j + M)T_s - \Delta t_j} \right] \\
&= \prod_{\substack{j \neq n \\ j=1}}^{N} \left[\frac{t - jT_s - \Delta t_j}{(n - j)T_s + \Delta t_n - \Delta t_j} \right] = P_n(t).
\end{aligned}
\tag{5.39}
$$

That allows to conclude that the orthogonal polynomials $P_n(t)$ in the available time window are exactly equal for an M time shift. This means that after M time instants, the Nth order orthogonal polynomial coefficients need not be computed all over again as they coincide with previous calculations. Thus considerable simplifications in complexity can be obtained. In addition, we know a priori the estimated values $\hat{\Delta t}_n$, $\forall n$, and we can therefore compute the MN different polynomial coefficients and store them in memory. Hence, we can think of $P_n(t) = P_n(m)$ for $m = 1, \ldots, M$. This leads to a time-varying filter implementation. The coefficients of the resulting time-variant filter are known, so the computational complexity is not really a big issue. However, it could also be implemented in parallel form, to further improve the performance. Figure 5.15 shows the compensation performance on a 12 bit 8 channel TI-ADC for a wideband QAM modulated signal using off-line mismatch error estimation and the simplified form of the Lagrange interpolator. There, it can be seen that a significant improvement can be obtained even for large mismatch errors in the order of 10%.

5.4   Additional Considerations: Training Signals for Measurement-Based Post-compensation

To perform the ADC post-compensation using measurement data, it is first required to obtain input–output measurements with input signals that allow a complete excitation of the acquisition system dynamics. In this manner, it is possible to

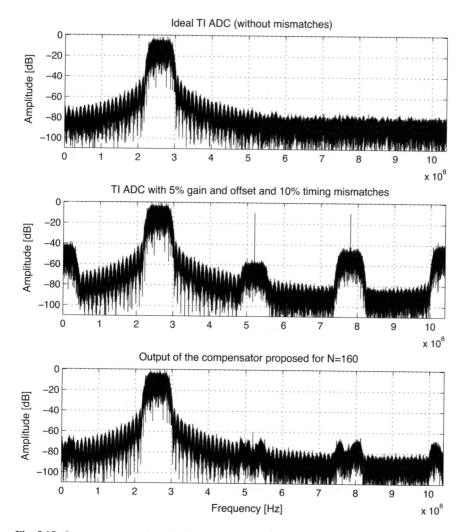

Fig. 5.15 Output spectrum of an ideal TI ADC, an uncompensated TI ADC with 5% gain and offset and 10% timing mismatches, and the output of the compensator proposed for $N = 160$ with 5% gain and offset and 10% timing mismatches

analyze the nonlinear effects of the device over several working ranges. It is also important to cover the complete dynamic range of the converter. Multitone input signals are considered to that purpose. As was shown earlier in this chapter for the case of sigma-delta converters, using this type of training signal allows for better linearization over a wider frequency range. However, even though some modern function generators are capable of generating multitone signals, the achievable SINAD levels are usually around 35–40 dB (equivalent to an ENOB $\cong 6$ bits) [14]. When considering ADCs with higher resolution, this SINAD levels are insufficient to obtain any performance improvement. In addition, a multitone test signal is problematic in practice, because the inter-modulation distortion (IMD) of the generator cannot be filtered and therefore the spectral purity of such signal cannot be enhanced. Furthermore, even single-tone sinusoids generated by most available function generators have lower SINAD levels than those required for compensation of high-resolution ADCs.

A possible solution to this issue is to enhance the spectral purity of sinusoids using filtering techniques and combine them into a novel training signal (a sequence of stepped sinusoids that do not add up in time but are concatenated) with spectral components similar to those of an equivalent multitone signal. To that purpose, we first consider the expression of a time-limited sinusoid input signal

$$f(t) = \cos(2\pi f_0 t)\,[u(t - t_{\mathrm{i}}) - u(t - t_{\mathrm{f}})] = g_1(t)[g_2(t)], \qquad (5.40)$$

where $u(t)$ is the Heaviside step function, t_{i} is the initial time instant, and t_{f} is the final instant. In (5.40), the term between brackets defines a time window $g_2(t)$ inside which the signal takes the cosine value $g_1(t)$, such that $f(t)$ is zero for $t < t_{\mathrm{i}}$ and $t > t_{\mathrm{f}}$. The Fourier transform of the time signal $f(t)$ in (5.40) is the frequency domain convolution of the transform functions corresponding to $g_1(t)$ and $g_2(t)$, which are multiplied in the time domain. On one hand, the transform of $g_1(t) = \cos(2\pi f t)$ is

$$G_1(f) = 1/2[\delta(f - f_0) + \delta(f + f_0)] \qquad (5.41)$$

while on the other hand, $g_2(t)$ in (5.40) can be alternatively expressed as a rectangular function $g_3(t)$ with a time shift $t_0 = (t_{\mathrm{f}} + t_{\mathrm{i}})/2$ and duration $\Delta t = (t_{\mathrm{f}} - t_{\mathrm{i}})$, with Fourier transform

$$G_3(f) = \mathrm{sinc}(f) \qquad (5.42)$$

such that the transform of $g_2(t)$ becomes

$$G_2(f) = e^{-j2\pi t_0 f}\frac{\Delta t}{2}\,\mathrm{sinc}\left(\frac{\Delta t f}{2}\right), \qquad (5.43)$$

where t_0 is a phase shift in the frequency domain and Δt determines the width of the sinc function main lobe, which is narrower for larger Δt. Thus, the Fourier transform of $f(t)$ in (5.40) is

$$F(f) = 1/2[\delta(f - f_0) + \delta(f + f_0)] \times e^{-j2\pi t_0 f}\frac{\Delta t}{2}\,\mathrm{sinc}\left(\frac{\Delta t f}{2}\right). \qquad (5.44)$$

Note that if $t_0 = 0$ and $\Delta t \to \infty$, then the sinc in (5.44) becomes an impulse and (5.44) reduces to (5.41). Finally, a time-limited multitone signal composed of N tones within the same time window can be posed as:

$$f_N(t) = \left[\sum_{n=1}^{N} \cos(2\pi f_n t) \right] [u(t - t_i) - u(t - t_f)]. \tag{5.45}$$

With the corresponding Fourier transform

$$F_N(f) = \frac{1}{2} \left[\sum_{n=1}^{N} [\delta(f - f_n) + \delta(f + f_n)] \right] \\ \times e^{-j2\pi t_0 f} \frac{\Delta t}{2} \operatorname{sinc}\left(\frac{\Delta t f}{2} \right). \tag{5.46}$$

We now consider a signal composed of several time-limited sinusoids that do not overlap over time, i.e., each sinusoid is multiplied by a different rectangular time window like the one described by (5.40), such that no overlapping in time occurs. In this case, we can express such signal in the time domain as

$$\tilde{f}_N(t) = \sum_{n=1}^{N} \{\cos(2\pi f_n t) [u(t - t_{in}) - u(t - t_{fn})]\} \tag{5.47}$$

with Fourier transform

$$\tilde{F}_N(f) = \frac{1}{2} \sum_{n=1}^{N} \left\{ [\delta(f - f_n) + \delta(f + f_n)] \\ \times e^{-j2\pi t_{on} f} \frac{\Delta t_{on}}{2} \operatorname{sinc}\left(\frac{\Delta t_{on} f}{2} \right) \right\}, \tag{5.48}$$

where $t_{on} = (t_{fn} + t_{in})/2$ is the time shift of the window corresponding to tone n and $\Delta t_{on} = (t_{fn} - t_{in})$ is the time duration of the associated window.

From (5.47) and (5.48), it can be seen that even though the sinusoid tones do not occur simultaneously in the time domain, the frequency components found in the spectrum of the signal [see (5.48)] are similar to those obtained using the multitone signal described in (5.45). While the ADC operates in the time domain and will only see one frequency at a time, the estimation of the compensator parameters is performed off-line with the full training sequence including all frequency components, and thus the estimated compensator is able to remove nonlinearities in a wider frequency range.

5.5   Summary of Key Points

In this chapter, the main ADC sampling architectures and structures are described at a functional level, making focus on the main features and characteristics they present. Next, traditional narrowband compensation techniques are introduced and briefly discussed: INL and DNL models, look-up tables, and dithering. Then, the problem of nonlinear distortion mitigation for wideband sampling is discussed in detail. When sampling wideband signals, especially at high rates and resolution, distortion is frequency dependent and shows very complex dynamics. Therefore, compensation of high-performance ADCs requires the use of nonlinear dynamic models capable to fully capture the characteristics of the device. We show that in these cases Hammerstein or Wiener models are insufficient, but more general Volterra models offer excellent compensation performance, as experimentally verified through measurements on high-performance commercial devices. In particular, parallel Wiener-like structures are found to provide an excellent trade-off between accuracy, generalization properties, and complexity. Time-interleaved structures are key to achieve very high sampling rates, as they allow to increase conversion speed beyond that of a simple ADC architecture. However, mismatches between individual ADCs can severely deteriorate performance in terms of effective resolution, and thus compensation is needed. A detailed analysis and description of mismatch errors is then provided, as well as background and foreground compensation techniques. Finally, some considerations on training signals for ADC post-compensation are briefly discussed.

References

1. R. Van de Plassche, *CMOS Integrated Analog-to-Digital and Digital-to-Analog Converters* (Kluwer Academic, Dordrecht, 2003)
2. A. Buchwald, High-speed time interleaved ADCs. IEEE Commun. Mag. **54**(4), 71–77 (2016)
3. L.D. Vito, H. Lundin, S. Rapuano, Bayesian calibration of a lookup table for ADC error correction. IEEE Trans. Instrum. Meas. **56**(3), 873–878 (2007)
4. F.H. Irons, D.M. Hummels, S.P. Kennedy, Improved compensation for analog-to-digital converters. IEEE Trans. Circuits Syst. **38**(8), 958–961 (1991)
5. C. Huang, H. Ting, S. Chang, Analysis of nonideal behaviors based on INL/DNL plots for SAR ADCs. IEEE Trans. Instrum. Meas. **65**(8), 1804–1817 (2016)
6. S. Medawar, B. Murmann, P. H′andel, N. Bj′orsell, M. Jansson, Static integral nonlinearity modeling and calibration of measured and synthetic pipeline analog-to-digital converters. IEEE Trans. Instrum. Meas. **63**(3), 502–511 (2014)
7. N. Bjorsell, Modeling Analog to Digital Converters at Radio Frequency, Doctoral Thesis in Telecommunications, Stockholm, Sweden (2007)
8. ADSP-2148X, One Technology Way, P.O. Box 9106, Norwood, MA 02062-9106 U.S.A. Analog Devices Inc. (2010)
9. H.F. Lundin, Characterization and Correction of Analog-to-digital Converters, Ph.D. Thesis, KTH, Stockholm, Sweden (2005)

10. S. Medawar, P. Handel, N. Bjorsell, M. Jansson, Postcorrection of pipelined analog-digital converters based on input-dependent integral nonlinearity modeling. IEEE Trans. Instrum. Meas. **60**(10), 3342–3350 (2011)
11. S. Ponnuru, M. Seo, U. Madhow, M. Rodwell, Joint mismatch and channel compensation for high-speed OFDM receivers with time-interleaved ADCs. IEEE Trans. Comm. **58**(8), 2391–2401 (2010)
12. P. Benabes, C. Lelandais-Perrault, N.L. Dortz, Mismatch calibration methods for high-speed time-interleaved ADCs, in *2014 IEEE 12th International New Circuits and Systems Conference (NEWCAS)* (2014), pp. 49–52
13. C.A. Schmidt, J.E. Cousseau, J.L. Figueroa, B.T. Reyes, M.R. Hueda, Efficient estimation and correction of mismatch errors in time-interleaved ADCs. IEEE Trans. Instrum. Meas. **65**(2), 243–254 (2016)
14. C.A Schmidt, O. Lifschitz, J.E. Cousseau, J.L. Figueroa, P. Julian, Methodology and measurement setup for analog-to-digital converter postcompensation. IEEE Trans. Instrum. Meas. **63**(3), 658–666 (2014)
15. J. Elbornsson, F. Gustafsson, J. Eklund, Blind equalization of time errors in a time-interleaved ADC system. IEEE Trans. Signal Process. **53**(4), 1413–1424 (2005)
16. Y. Qiu, Y.J. Liu, J. Zhou, G. Zhang, D. Chen, N. Du, All-digital blind background calibration technique for any channel time-interleaved ADC. IEEE Trans. Circuits Syst. Regul. Pap. **PP**(99), 1–12 (2018)
17. C. Schmidt, J.E. Cousseau, J.L. Figueroa, R. Wichman, S. Werner, Characterization and compensation of nonlinearities in a continuous-time first-order ADC, in *2010 IEEE International Microwave Workshop Series on RF Front-ends for Software Defined and Cognitive Radio Solutions (IMWS)* (2010), pp. 1–4
18. C.A. Schmidt, J.E. Cousseau, J.L. Figueroa, R. Wichman, S. Werner, Non-linearities modelling and post-compensation in continuous-time $\sigma\delta$ modulators. *IET Microwaves Antennas Propag.* **5**(15), 1796–1804 (2011)
19. C.A. Schmidt, J.L. Figueroa, J.E. Cousseau, A.M. Tonello, Pilot-based TI-ADC mismatch error calibration for IR-UWB receivers. IEEE Access **7**, 74340–74350 (2019)
20. M. Schetzen, *The Volterra and Wiener theories of Nonlinear Systems* (John Wiley and Sons Inc., New York, 1980)
21. F.J. Doyle, R.K. Pearson, *Identification and Control Using Volterra Models* (Springer, London, 2002)
22. S. Boyd, L.O. Chua, Fading memory and the problem of approximating nonlinear operators with Volterra series. IEEE Trans. Circuits Syst. **32**(11), 1150–1161 (1985)
23. I. Ryan, H. Mahdi, An oversampled rate converter using sigma delta noise shaping, in *IET Irish Signals and Systems Conference*, Dublin (2009), pp. 1–6
24. E. Bonizzoni, A.P. Perez, F. Maloberti, M. Garcia-Andrade, Third-order $\sigma - \delta$ modulator with 61-db SNR and 6-MHz bandwidth consuming 6 mW, in *34th European Solid-State Circuits Conference*, Edinburgh (2008), pp. 218–221
25. T-S. Jeong, W. Choi, J. Gi, C. Yoo, Low voltage analog digital converter using sigma-delta modulator, in *International SoC Design Conference*, Busan (2008), pp. III52–III53
26. W-L. Yang, W-H. Hsieh, C-C. Hung, A third-order continuous-time sigma-delta modulator for Bluetooth, in *International Symposium on VLSI Design, Automation and Test*, Hsinchu (2009), pp. 247–250
27. A. Morgado, R. del Rio, J.M. de la Rosa, F. Medeiro, B. Perez-Verdu, F.V. Fernandez, A. Rodriguez-Vazquez, Reconfiguration of cascade sigma delta modulators for multistandard GSM/Bluetooth/UMTS/WLAN transceivers, in *IEEE International Symposium on Circuits and Systems*, Island of Kos (2006), pp. 1884–1887
28. B.R. Jose, P. Mythili, J. Singh, J. Mathew, A triple-mode sigma-delta modulator design for wireless standards, in *10th International Conference on Information Technology*, Orissa (2007), pp. 17–20

29. A. Leuciuc, On the nonlinearity of integrators in continuous-time delta-sigma modulators, in *IEEE Midwest Symposium on Circuits and Systems*, Dayton (2001), pp. 862–865
30. S. Pavan, Efficient simulation of weak nonlinearities in continuous-time oversampling converters. IEEE Trans. Circuits Syst. **57**(8), 1925–1934 (2010)
31. P. Sankar, S. Pavan, Analysis of integrator nonlinearity in a class of continuous-time delta-sigma modulators. IEEE Trans. Circuits Syst. **54**(12), 1150–1161 (2007)
32. T. Karema, T. Ritoniemi, H. Tenhunen, Intermodulation in sigma-delta D/A converters, in *IEEE International Symposium on Circuits and Systems*, Montreal (1991), pp. 1625–1628
33. M. Keramat, Functionality of quantization noise in sigma-delta modulators, in *IEEE Midwest Symposium on Circuits and Systems*, Lansing (2000), pp. 912–915
34. L. Samid, Y. Manoli, A multibit continuous time sigma delta modulator with successive-approximation quantizer, in *IEEE International Symposium on Circuits and Systems*, Island of Kos (2006), pp. 2965–2968
35. D.R. Morgan, Z. Ma, J. Kim, M.G. Zierdt, J. Pastalan, A generalized memory polynomial model for digital predistortion of RF power amplifiers. IEEE Trans. Circuits Syst. **54**(10), 3852–3860 (2006)
36. J.C. Gómez, E. Baeyens, Identification of block-oriented nonlinear systems using orthonormal bases. J. Process Control **14**(6), 685–697 (2003)
37. P. Nikaeen, B. Murmann, Digital compensation of dynamic acquisition errors at the front-end of high-performance A/D converters. IEEE J. Sel. Top. Sign. Proces. **3**(3), 499–508 (2009)
38. ADSP-2148X, One Technology Way, P.O. Box 9106, Norwood, MA 02062-9106 U.S.A. Analog Devices Inc. (2010)
39. F. Centurelli, P. Monsurr, F. Rosato, D. Ruscio, A. Trifiletti, Calibration of pipeline ADC with pruned Volterra kernels. Electron. Lett. **52**(16), 1370–1371 (2016)
40. S.M.R. Islam, N. Avazov, O.A. Dobre, K.S. Kwak, Power-domain non-orthogonal multiple access (NOMA) in 5G systems: potentials and challenges. IEEE Commun. Surv. Tutorials **PP**(99), 1–1 (2016)
41. B.T. Reyes, R.M. Sanchez, A.L. Pola, M.R. Hueda, Design and experimental evaluation of a time-interleaved ADC calibration algorithm for application in high-speed communication systems. IEEE Trans. Circuits Syst. Regul. Pap. **64**(5), 1019–1030 (2017)
42. C.R. Anderson, S. Venkatesh, J.E. Ibrahim, R.M. Buehrer, J.H. Reed, Analysis and implementation of a time-interleaved ADC array for a software-defined UWB receiver. IEEE Trans. Veh. Technol. **58**(8), 4046–4063 (2009)
43. V.T.D. Huynh, N. Noels, H. Steendam, Offset mismatch calibration for TI-ADCs in high-speed OFDM systems, in *2015 IEEE Symposium on Communications and Vehicular Technology in the Benelux (SCVT)* (2015), pp. 1–5
44. S.K. Sindhi, K.M.M. Prabhu, Reconstruction of N-th order nonuniformly sampled bandlimited signals using digital filter banks. Digital Signal Process. **23**, 1877–1886 (2013)
45. J. Elbornsson, F. Gustafsson, J.E. Eklund, Blind adaptive equalization of mismatch errors in a time-interleaved A/D converter system. IEEE Trans. Circuits Syst. Regul. Pap. **51**(1), 151–158 (2004)
46. C. Vogel, H. Johansson, Time-interleaved analog-to-digital converters: status and future directions, in *2006 Proceedings of IEEE International Symposium on Circuits and Systems, 2006. ISCAS 2006* (2006), 3386–3389
47. C. Vogel, M. Hotz, S. Saleem, K. Hausmair, M. Soudan, A review on low-complexity structures and algorithms for the correction of mismatch errors in time-interleaved ADCs, in *2012 IEEE 10th International New Circuits and Systems Conference (NEWCAS)* (2012), pp. 349–352
48. S. Chen, L. Wang, H. Zhang, R. Murugesu, D. Dunwell, A.C. Carusone, All-digital calibration of timing mismatch error in time-interleaved analog-to-digital converters. IEEE Trans. Very Large Scale Integr. VLSI Syst. **25**(9), 2552–2560 (2017)
49. S. Kwon, S. Lee, J. Kim, A joint timing synchronization, channel estimation, and SFD detection for IR-UWB systems. J. Commun. Networks **14**(5), 501–509 (2012)
50. P.P. Vaidyanathan, *Multirate Systems and Filter Banks* (Prentice Hall PTR, Englewood Cliffs, 1993)

51. T.I. Laakso, V. Valimaki, M. Karjalainen, U.K. Laine, Splitting the unit delay [FIR/all pass filters design]. IEEE Signal Process. Mag. **13**(1), 30–60 (1996)
52. G. Qin, G. Liu, M. Gao, X. Fu, P. Xu, Correction of sample-time error for time-interleaved sampling system using cubic spline interpolation. Metrol. Measur. Syst. **21**(3), 485–496 (2014)

Chapter 6
Frequency Offset and Phase Noise

Abstract Multicarrier systems are widely used due to their high spectral efficiency and robustness against frequency-selective channels. However, the performance of these systems is highly affected by carrier frequency offset (CFO) and phase noise (PN), since the orthogonality between carriers is lost. As these imperfections are mostly related to local oscillator quality and user mobility, low-cost and high-speed applications are critical. The estimation of CFO and phase noise is divided into two steps, acquisition and tracking, where known sequences (or pilot symbols) are used to estimate the parameters at the receiver. The multiuser case is more challenging since the problem is multi-parametric, i.e., it is necessary to estimate the CFO and phase noise of each user. While the compensation of these imperfections is quite simple in the single user case (downlink), it is more elaborated in a multiuser access condition (uplink). Successive interference cancellation or linear suppression techniques are therefore used for CFO and phase noise cancellation in the multiuser case. In this chapter we first describe the effects of the CFO and the phase noise in the system performance, paying special attention to critical applications, such as low-cost unstable oscillators and high-speed vehicles. Then, we introduce several CFO estimation techniques for the downlink (single user case), based on statistical properties of the sequence of pilot symbols. Finally, we present estimation and compensation strategies for the uplink (multiuser case) of orthogonal frequency division multiple access (OFDMA) and multiuser filter bank multicarrier (FBMC), using direct and successive interference cancellation algorithms. The performance of the compensation is evaluated considering high and low mobility scenarios.

6.1 Effects of the CFO and Phase Noise in the System Performance

Synchronization plays an important role in the design of digital communication systems. Essentially this function is responsible for recovering some reference parameters from the received signal, which are necessary for a reliable symbol

© Springer Nature Switzerland AG 2020
F. Gregorio et al., *Signal Processing Techniques for Power Efficient Wireless Communication Systems*, Signals and Communication Technology,
https://doi.org/10.1007/978-3-030-32437-7_6

detection. The most relevant are sample clock, time, and frequency synchronization. In modern systems, current technology ensures that the difference between sample clocks is low enough as not to degrade system performance in a noticeable manner, so that in general this effect is not considered. An exception to this are direct-RF-sampling receivers, where bandpass sampling is used to reduce the implementation complexity [1]. Time synchronization refers to the estimation of the beginning of the frame and the identification of each multicarrier symbol, i.e., the DFT-window positioning. The CP samples that are not corrupted by the channel can be used to make the system more robust to time synchronization errors. This scenario is called quasi-synchronous scenario. Then, time synchronization is not critical for CP-added modulations like OFDM(A). On the other hand, frequency synchronization aims to minimize the difference between the carrier frequency of the received signal and the frequency of the local oscillator. The lack of frequency synchronism produces a loss of orthogonality between subcarriers which reduces system performance, as it introduces intercarrier interference (ICI) leading to a degradation in bit error rate (BER).

Frequency synchronization errors produce carrier frequency offset (CFO), and may have two sources: the difference between the frequency of the transmitter and receiver oscillators, and the Doppler effect due to the relative motion between transmitter and receiver. Another impairment related with CFO is the phase noise (PN), caused by random fluctuations in the oscillator.

6.1.1 Critical Applications: High Frequency Oscillators and High-Speed Vehicles

As the CFO and PN are produced by different physical phenomena, the magnitude and time evolution of these imperfections could be different. In the following, CFO and PN models to describe their behavior are introduced.

The CFO may have two main sources: f_{lo} produced by manufacturing deviations in the components, mounting defects, temperature, etc., and f_d due to the Doppler effect caused by mobility. To give an idea of the relative magnitudes, f_{lo} is in the order of ± 10 ppm of the carrier frequency [2], whereas the maximum Doppler shift even for high-speed mobiles is several times smaller. For example, for a speed of 200 km/h the equivalent Doppler shift is 0.185 ppm.

The term f_{lo} can be considered static in practice, since it depends on either fixed or slowly varying parameters. On the other hand, f_d varies, although the variation is slow even for high speeds [3, 4]. Only channels with asymmetric Doppler spectrum produce CFO. As this scenario typically corresponds to channels with a line-of-sight component, Rician channel models are the most suitable. The scenario is sketched in Fig. 6.1, where a mobile with speed v moves relative to the base station following

Fig. 6.1 Dynamic of the
mobile for modeling of the
time-varying CFO

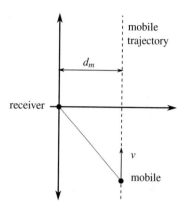

a straight path, producing a variation in the phase of the line-of-sight component [3].
Considering channel tap variances $\{\sigma_l^2\}_{l=0}^{N_h-1}$, the average CFO shift of the symbol ℓ
is given by

$$\bar{f}_d(\ell) = \frac{\sigma_0^2 K_r f_d(\ell)}{(K_r + 1) \sum_{l=0}^{N_h-1} \sigma_l^2}, \tag{6.1}$$

where K_r is the Rice factor, $f_d(\ell)$ is the Doppler shift of the line-of-sight component
for the OFDM symbol ℓ. Following a linear approximation, the variations in the
Doppler frequency can be described as:

$$f_d(\ell) = -f_c \frac{v^2}{cd_m} T_s \ell, \tag{6.2}$$

where d_m is the distance between the transmitter and the receiver and T_s is the
OFDM symbol length. As a consequence, it is possible to model the variant CFO as:

$$f(\ell) = f_{lo} + \bar{f}_d(\ell), \tag{6.3}$$

where f_{lo} is constant and represents the carrier frequency shift due to oscillator
inaccuracies and $\bar{f}_d(\ell)$ is time-variant and corresponds to the shift due to Doppler
effect.

Random fluctuations in the oscillator phase due to inherent noise in the electronics of the device produce PN, as introduced in Sect. 2.8. The effect is not large for
carrier frequencies lower than 5 GHz, considering current oscillators technology.
On the other hand, for higher carrier frequencies, as those used in mm-wave
communications, the PN needs to be compensated.

6.1.2 Effects of Carrier Frequency Offset and Phase Noise in OFDM

Although the sources of CFO and PN are different, they can be included in a single OFDM frequency domain model. From (2.26) in Sect. 2.4, the received signal can be written as:

$$\mathbf{y} = \mathbf{\Pi H x} + \mathbf{w}, \tag{6.4}$$

where $\mathbf{H}$, $\mathbf{x}$, and $\mathbf{w}$ are, respectively, the circulant channel matrix, the transmitted symbol vector, the AWGN vector with covariance matrix $\sigma^2 \mathbf{I}_N$, and $\mathbf{I}_N$ the identity matrix of $N \times N$. The matrix $\mathbf{\Pi}$ is a circulant CFO and PN matrix with its first column defined as

$$\Pi(k) = \frac{1}{N} \sum_{n=0}^{N-1} \exp\left(j\upsilon(n) + \frac{j2\pi\xi n}{N}\right) \exp\left(\frac{-j2\pi nk}{N}\right), k = 0, \ldots, N-1. \tag{6.5}$$

The term $\xi = f(\ell)T_s$ is the CFO normalized to the intercarrier spacing, whereas $\upsilon(n)$ is a random variable that models the receiver PN [5]. Note that ξ can be considered constant within the OFDM symbol ℓ. It should be noted that the same model can also be used to include the effect of transmitter PN, if the oscillator has small phase increment during an OFDM symbol. In that case, $\upsilon(n)$ is the sum of the contributions of each oscillator [5].

To analyze the effects of frequency errors, an expression of the received symbol at carrier k is included

$$Y(k) = \Pi(0)H(k)X(k) + \sum_{l=0, l\neq k}^{N-1} \Pi(k-l)H(l)X(l) + W(k). \tag{6.6}$$

The term $\Pi(0)$ is common to all the subcarriers and is called common phase error (CPE), whereas the term of the sum describes the ICI. In order to avoid significant performance loss, the frequency error should be maintained lower than 1 or 2% of the intercarrier spacing. In this context, the LTE standard fixes the maximum frequency shift within a frame to 0.1 ppm.

Let us analyze separately the effect of the CFO and the phase noise. For CFO only, i.e., $\upsilon(n) = 0$ in (6.5), multiplication by $\mathbf{\Pi}$ produces a frequency shift in the received spectrum, as can be seen in Fig. 6.2 [4]. Additionally, the CPE and the ICI contributions are pointed out in the curves. On the other hand, the isolated effect of the PN, i.e., $\xi = 0$ in (6.5), is shown in Fig. 6.3. Two scenarios are considered, low and large phase variation (low or large PN spectral density bandwidth). Figure 6.3a shows that for low PN variation, the CPE effect is dominant. On the other hand, from Fig. 6.3b it is evident that the ICI is more notorious for large PN variations [6].

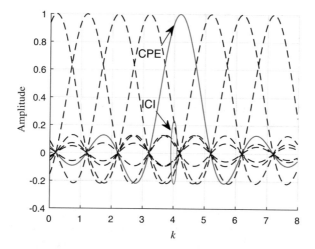

Fig. 6.2 CFO effect over an OFDM signal in frequency domain

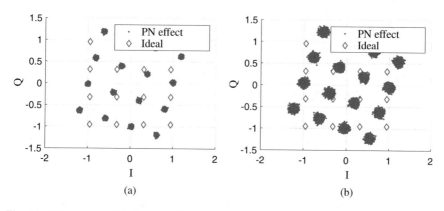

Fig. 6.3 Effect of the PN in the constellation. In (**a**) is shown a PN of 5 Hz of bandwidth, whereas in (**b**) one of 50 Hz

For a user with low mobility, the CFO is almost constant and its estimation is required once, or at most a few times during transmission. However, if the user has high mobility, the CFO can change and its estimation can be required several times during communication. This scenario is studied in Fig. 6.13 later on. Considering current technology, oscillators are not stable enough to warranty the operation in mm-wave bands without PN compensation. Given the random nature of the PN, the estimation has to be done for each received symbol. For moderate PN levels, as those shown in Fig. 6.3a, this is not critical in terms of complexity since the CPE is a single complex parameter easy to estimate. For extreme cases, as depicted in Fig. 6.3b, the ICI part also needs to be compensated. This increments considerably the computational complexity since all the samples of the PN process $\upsilon(n)$ have to be estimated for each received symbol. This condition is analyzed in Fig. 6.6 of Sect. 6.3.

6.2 Estimation Techniques for the Downlink (Single User Case)

The synchronization process is typically separated into acquisition and tracking. During acquisition, training sequences with repetitive structures are used to obtain initial estimations of the synchronization parameters [7–10]. As in this phase, time and frequency scales are not yet aligned with the received sequence, algorithms must tolerate large synchronization errors. The tracking stage refines initial estimations as well as it estimates small variations of the local oscillator and Doppler shift. To this purpose, the redundancy introduced by the CP or pilot symbols inserted between data symbols can be used [11, 12].

Communication standards as WiMax [13] or LTE [14] separate data transmissions in frames as depicted in Fig. 6.4, where a reference sequence is included to assist the synchronization process.

The block without data located at the beginning of the frame can be used to estimate the noise power and interference. Additionally it provides a simple method to estimate the beginning of the frame.

After frame detection and timing synchronization, each terminal must align its local oscillator to the carrier frequency of the received signal. This operation is known as frequency acquisition and is usually performed for each received frame or according to channel condition (CFO variability). CFO estimation is a nonlinear problem in the parameters and can be generally classified into data-aided, and non-data-aided or blind estimators. The former estimators use a training sequence which is typically included in the preamble or symbol modulated in dedicated carriers called pilots [7, 9, 15]. The latter estimators refer to those that exploit the statistical structure of the received signal to perform the estimation.

As the communication channel is not known at this point, symbols of the training sequence cannot be considered known. However, both estimations are decoupled, this implies that it is possible to estimate and compensate for the CFO first, and then to estimate the channel [9]. The repetitive structure of the training sequence is not

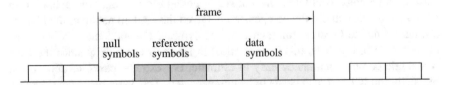

Fig. 6.4 Data frame structure

lost when the signal passes through the channel, except for a phase shift caused by the CFO. This allows to find CFO estimators without channel knowledge, based on this phase shift.

In the following, a brief overview of state-of-the-art CFO estimators for OFDM systems based on repetitive sequences is presented. The maximum likelihood estimator (MLE) of CFO is presented in [9]. Besides its theoretical importance, this solution is not used in practice due to the high associated computational complexity (grid search). Low-complexity solutions typically follow three basic steps: an indirect measure of the CFO (usually the phase of the autocorrelation), an ambiguity reduction technique to increase the estimation range, and finally a combination of the partial estimates (for repetitive sequences of more than two periods).

To introduce the methods, (6.4) without PN ($v(n) = 0$) is rewritten in time domain as follows:

$$r(n) = e^{j\frac{2\pi\xi}{N}n}q(n) + w(n), \text{ for } n = 0, \ldots, N-1, \tag{6.7}$$

where $q(n) = \sum_{l=0}^{N_h-1} h(l)p(n-l)$ is an M-periodic sequence of length N with $J = N/M$ periods, and the transmitted signal is redefined as a training sequence $p(n)$, $\{k - L_{cp} \leq n \leq N-1\}$, also of period M and length $N + L_{cp}$. Note that (6.7) is the received signal (2.14) with CFO. Then, the autocorrelation of the received sequence results

$$\hat{\Gamma}(k) = \frac{1}{N-kM} \sum_{n=kM}^{N-1} r(n)r^*(n-kM), \quad 1 \leq k \leq J-1. \tag{6.8}$$

One of the first proposals was Moose's algorithm [8], which employs two identical training sequences to perform the CFO estimation. The main drawback of this method is the low acquisition range that is less than half of the intercarrier spacing. Another estimation method was proposed by Schmidl and Cox [7], which also employs two training sequences but they are designed to ensure a full acquisition range, i.e., $|\hat{\xi}| \leq N/2|$. The first sequence has two identical halves and is the same that is employed for timing acquisition. The second has a pseudo-random sequence in even carriers and another in odd carriers. In this method, CFO is separated in a fractional part ξ_f, between $(-1, 1)$, and an integer part ξ_{en}, then the CFO can be written as $\xi = \xi_f + 2\xi_{en}$. The estimation is obtained from $\hat{\Gamma}(J/2)$, using training sequences of $J = 2$ periods. Besides the noise, the two halves are identical except for a phase shift of $\pi\xi_f$. Then, the CFO estimation can be obtained as

$$\hat{\xi}_f = \frac{1}{\pi} \arg\{\hat{\Gamma}(J/2)\}. \tag{6.9}$$

With the estimation $\hat{\xi}_f$, the fractional part is compensated by counter-rotating the received sequence at an angular sequence $2\pi\hat{\xi}_f/N$. The term $2\xi_{en}$ represents a circular shift of the symbols in the carriers, but does not introduce ICI. To estimate the integer part, the second training block is used [7].

An extension of Schmidl and Cox algorithm for training sequences with $J > 2$ was proposed by Morelli and Mengalli [16]. Replacing (6.7) in (6.8), the autocorrelation can be written as:

$$\hat{\Gamma}(k) = e^{\frac{j2\pi \xi k}{J}} \chi(k), \ 1 \le k \le J - 1, \tag{6.10}$$

where $\chi(k)$ is a sequence depending on the training sequence, the channel, and the noise. It is easy to note that CFO information is divided into $J - 1$ components of $\hat{\Gamma}(k)$. Defining

$$\theta(k) = \arg\{\hat{\Gamma}(k)\} = \frac{2\pi \xi k}{J} + \arg\{\chi(k)\}, \ 1 \le k \le J - 1, \tag{6.11}$$

and considering that $|\xi| < J/(2k)$, an estimation of ξ can be obtained. As the estimation range varies with k, it is not possible to directly combine the available information without reducing the estimation range. To avoid the range reduction, the algorithm of Morelli bases its estimation in the phase difference of $\hat{\Gamma}(k)$, which is defined as

$$\theta_d(k) = [\theta(k) - \theta(k-1)]_{2\pi} = \left[\frac{2\pi \xi}{J} + \gamma(k)\right]_{2\pi}, \ 1 \le k \le A, \tag{6.12}$$

where $[\cdot]_{2\pi}$ denotes reduction to the interval $[0, 2\pi)$, $\gamma(k) = \arg\{\chi(k)\} - \arg\{\chi(k-1)\}$ is a zero mean noise, $1 \le A \le J - 1$ and $\theta(0) = 0$. As can be concluded from (6.12), the estimation range is now $|\xi| < J/2$. Finally, grouping the phases in the vector $\boldsymbol{\theta}_d = [\theta_d(1), \cdots, \theta_d(A)]^T$ and assuming a high SNR, the best linear unbiased estimator (BLUE) to obtain the CFO estimation results

$$\hat{\xi}_M = \mathcal{B}\{\boldsymbol{\theta}_d, \mathbf{C}_\theta\}, \tag{6.13}$$

where

$$\mathcal{B}\{\boldsymbol{\theta}, \mathbf{C}\} = \frac{J}{2\pi} \frac{\mathbf{1}^T \mathbf{C}^{-1} \boldsymbol{\theta}}{\mathbf{1}^T \mathbf{C}^{-1} \mathbf{1}}, \tag{6.14}$$

$\mathbf{C}$ is the covariance matrix of the sample vector $\boldsymbol{\theta}$, and $\mathbf{1}$ has the same number of rows than $\mathbf{C}$. Since high-order noise terms are disregarded in the derivation of the covariance matrix of $\theta_d(k)$, it becomes singular for $k > J/2$ and it is not possible to employ the information of autocorrelation lags larger than $J/2$ [16]. The best performance is obtained for $A = J/2$ [9].

On the other hand, Minn proposed another algorithm in [17] based on the function

$$\zeta(k) = \frac{J}{2\pi k}\theta(k) = \xi + \frac{J}{2\pi k}\arg\{\chi(k)\}, \ 1 \leq k \leq J - 1. \tag{6.15}$$

To extend the range of the estimate, $\zeta(1)$ is first used as a coarse CFO estimate. Then, the received signal is compensated using $\zeta(1)$ to obtain $\tilde{y}(n) = \exp(-j2\pi\zeta(1)n/N)y(n)$. Replacing $y(n)$ in (6.8) by $\tilde{y}(n)$, the angles $\tilde{\zeta}(k)$ can be obtained from (6.15), corresponding to the residual CFO, i.e., $\zeta(k) = \zeta(1)+\tilde{\zeta}(k)$ for $2 \leq k \leq J-1$. Finally, grouping the phases in the vector $\boldsymbol{\zeta} = [\zeta(1), \cdots, \zeta(J-1)]^\mathsf{T}$ and assuming high SNR, the BLUE of the CFO is

$$\hat{\xi}_{Mi} = \mathscr{B}\{\boldsymbol{\zeta}, \mathbf{C}_\zeta\}. \tag{6.16}$$

Three covariance matrices $\mathbf{C}_\zeta$ are presented in [17], corresponding to different approaches. Two of them include high-order noise terms allowing the use of all available information ($J - 1$ phases), and the third one discards these terms leading to a singular covariance matrix for $k > J/2$, as in [16]. Since the coarse CFO estimate depends only on the first autocorrelation coefficient of $\zeta(k)$, the technique derived in [17] is not robust at low SNR.

The main conclusion after Minn's contribution is that information loss is not related with the methodology employed to avoid phase ambiguity, but with the approximations involved in the derivation of the correlation matrix. In other words, *if high-order noise terms are included in the covariance matrix used in the linear combination (BLUE), the resulting estimator is able to use all information in the training sequence ($J - 1$ autocorrelation coefficients).* Using this conclusion, the high-order noise terms (HONT) estimator is proposed in [18], which considers a correction in the derivation of the covariance matrix of Morelli's algorithm to use all available information to make the estimation. In this case, the CFO estimation results

$$\hat{\xi}_{HONT} = \mathscr{B}\{\boldsymbol{\theta}_d, \mathbf{C}_{HONT}\}. \tag{6.17}$$

Here, $\mathbf{C}_{HONT}$ is derived considering high-order noise terms so it allows the use of all available information. Then, the matrix has dimension $(J - 1) \times (J - 1)$, instead of $J/2 \times J/2$ as $\mathbf{C}_\theta$ in (6.13). HONT algorithm requires the inverse of the phase-difference covariance matrix to combine the information. In [18], some low-complexity solutions are presented, assuming independence between phase terms.

As a summary of the section, the performance of the presented algorithms is illustrated. Training sequences of 64 samples and period 8 are used in the simulation. The length of the cyclic prefix is 16, and the channel has 10 coefficients and exponential power decay profile.

Figure 6.5 illustrates the performance of HONT, MLE, and the estimators of Morelli and Minn in terms of mean square error (MSE) for $\xi = 0.3$. As expected,

Fig. 6.5 MSE versus SNR comparison for CFO estimation algorithms with $\xi = 0.3$

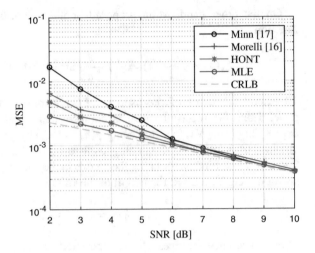

Fig. 6.6 PN compensation considering an oscillator of 1 kHz of bandwidth and a power of -15 dBc. CPE and CPE + ICI algorithms are included

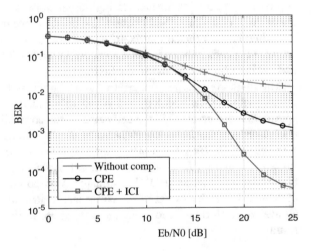

the MLE has the best performance. It is noted that HONT outperforms Morelli and Minn estimators at low SNR. At moderate SNR (higher than 4 dB), MLE, HONT, and Minn estimator have similar performance and outperform Morelli estimator. This is because Morelli estimator only employs $J/2$ autocorrelation phases. At low SNR (lower than 4 dB), Minn estimator loses robustness due to the coarse estimation described earlier in this section and its performance decreases considerably. All algorithms converge to the CRLB for high SNR (higher than 10 dB).

The PN estimation performance is shown in Fig. 6.6, where the performance of CPE and CPE + ICI algorithms proposed in [5] is compared with a non-compensated system. The oscillator bandwidth is 1 kHz with a power of -15 dBc. As it can be noted, the system performance is highly affected by the PN. The CPE algorithm is able to compensate for some of the interference but the ICI produces an error floor at BER 1×10^{-3}. The CPE+ICI algorithm improves significantly the performance. This compensation is considerably more complex than CPE since all the components of $\upsilon(n)$ have to be estimated. To make the complexity affordable, only 3 components to each side of the main diagonal of Π are considered. Many other algorithms that improve the performance of [5] have been proposed. In [19, 20] some improvements in the interpolation of the PN are introduced. For more harsh scenarios where the communication channel is not known, [21] propose a joint channel and PN estimation algorithm. A good summary of the state-of-the-art in PN modeling and compensation can be found in [22].

6.3 Estimation and Compensation Techniques for the Uplink (Multiuser Case)

In multiuser systems, each terminal synchronizes its local oscillator and time reference using downlink signals. Therefore, timing errors in the uplink are only due to propagation delays, and frequency synchronization errors due to oscillator drift and Doppler effect. If these errors are small, i.e., if the frequency shift due to Doppler is much lower than the intercarrier spacing, and the cyclic prefix length is long enough to cope with the channel delay spread and channel length, then estimation and compensation are not necessary. On the other hand, if these conditions cannot be satisfied, synchronization routines for the uplink need to be applied [23].

The multiuser synchronization context in the uplink is much more complex than the single user case (or downlink), because it is a multi-parametric problem. In other words, it is necessary to estimate and compensate for synchronization errors of each user. In general, used techniques depend on the chosen carrier allocation scheme (CAS) since this determines how easy is to separate the signals of each user. Naturally, user separation is trivial for subband CAS (SCAS) and more challenging for generalized CAS.

Frequency synchronization is a two-step procedure: CFO estimation and CFO compensation. CFO estimation relies on a repetitive structure of special symbols inserted at the beginning of the frame [23, 24]. When subband CAS is considered, the estimation problem is similar to the single user case, since user separation is easily achieved with a filter bank. For the interleaved CAS (ICAS) case, the inner structure generated by the allocation can be used to derive an estimator based on subspace methods [25]. The generalized GCAS case is the most complex and proposed solutions are based on the maximum likelihood estimator or iterative approximations as the expectation maximization method. The MSE of the CFO estimation for a system with GCAS, 128 subcarriers, a CP of 8 subcarriers, and 4 users is shown in Fig. 6.7. The figure shows that the algorithm is able to estimate the CFO very accurately [26].

Fig. 6.7 CFO estimation for
the OFDM uplink, for a
system with 128 subcarriers,
a CP of 8 subcarriers, and 4
users

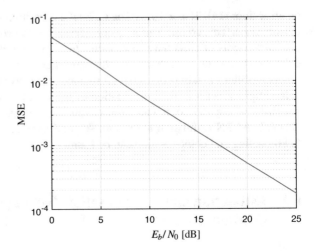

Once estimated, the frequency error must be corrected. The case of subband CAS
is trivial, since the signal of each user is available and the same phase counter-
rotation used in the previous section for the single user case can be applied [27]. For
the interleaved and generalized CAS, the task is more complex since the frequency
error produces ICI and users signals cannot be easily decoupled. In this cases, the
compensation is performed by applying interference cancellation [28–31] or linear
detection techniques [32]. Linear suppression outperforms interference cancellation
at the cost of higher complexity. That complexity is related to the inversion of
the matrix that describes the CFO interference. Therefore, banded approximations
of that matrix can be used to simplify the linear suppression at the expense of
some performance loss [32, 33]. Iterative approximations, that process each user
separately, attain a good performance for subband CAS [34]. If a linear suppression
stage for each user is added to the iterative cancellation, the result outperforms the
banded approximation, and has lower computational complexity than the complete
matrix inverse and the single iterative algorithm, for a system of few subcarriers
[35]. Nevertheless, the structure of the induced CFO interference is not used,
leaving some room for further complexity reduction. Another iterative approach
to compensate for the CFO is based on the Newton's method [36]. Despite the
great complexity reduction obtained, the method does not achieve good performance
for different multiuser contexts. Also, algorithms based on the conjugate gradient
method achieve a good compensation performance with low complexity, but the
particular structure of the interference matrix is again not fully employed to obtain
better computational efficiency [37]. Finally, to further reduce the complexity, the
interference structure can be taken into account by using the space-alternating
generalized expectation–maximization (SAGE) approach [38, 39].

The CFO also produces multiple access interference in FBMC, but that interference can be made negligible by leaving only a few guard subcarriers between
each user, thanks to the sharp cutoff frequency of the filters [40]. As a consequence,

only intercarrier interference needs to be compensated in FBMC. Therefore, a low-complexity compensation method for CFO in FBMC with subband CAS can be introduced [40]. The compensation is equivalent to the *post*-FFT compensation derived for OFDMA [27], although it performs better due to the low multiple access interference inherent to FBMC.

Even when a low-complexity polyphase realization exists [41], the high complexity of the whole FBMC system (i.e., offset QAM (OQAM) and channel estimation/equalization), including CFO estimation and compensation, is its main drawback if considered as alternative to efficient CFO-compensated OFDMA.

6.3.1 CFO Compensation for OFDMA

Multicarrier techniques are being considered for the uplink of wireless systems due to their high data throughput, spectral efficiency, and versatility [42, 43]. Orthogonal frequency division multiple access (OFDMA) is based on the inherited orthogonality of the OFDM modulation, where a subset of subcarriers are assigned to each user according to a CAS. In the uplink of centralized systems, the signals of different users are added together at the base station (BS) [44, 45]. The multicarrier symbol has N subcarriers, where $M < N$ subcarriers are used for data transmission and the remaining $(N - M)$ are virtual subcarriers (VS) located at the edges of the band. Virtual subcarriers avoid frequency leakage to the neighbor bands [46]. Useful subcarriers are divided into L subchannels, where each subchannel containing M/L subcarriers corresponds to a different user.

From the single user equation (6.4), it is possible to find an expression for the uplink (multiuser case) as follows. The receiver block $\mathbf{y}$ of size N at the BS, after cyclic prefix removal and the discrete Fourier transform (DFT), can be written as:

$$\mathbf{y} = \sum_{m=0}^{L-1} \mathbf{y}^{(m)} + \mathbf{w}, \qquad (6.18)$$

where $\mathbf{y}^{(m)} = \mathbf{\Pi}^{(m)} \mathbf{s}^{(m)}$ is the signal received from user m corrupted by the channel and the CFO, $\mathbf{\Pi}^{(m)}$ is the CFO interference matrix of user m as defined in (6.4), $\mathbf{s}^{(m)} = \mathbf{H}^{(m)} \mathbf{x}^{(m)}$, $\mathbf{x}^{(m)}$ is the symbol of user m, and $\mathbf{H}^{(m)}$ is the channel matrix between the user m and the base station.

The ideal received signal in frequency domain, i.e., without CFO, is given by $\mathbf{s} = \sum_{m=0}^{L-1} \mathbf{s}^{(m)}$. Considering the signal of each user $\mathbf{s}^{(m)}$, it is possible to write

$$\mathbf{s}^{(m)} = \mathbf{\Psi}^{(m)} \mathbf{s}, \qquad (6.19)$$

where $\boldsymbol{\Psi}^{(m)}$ is a diagonal selection matrix that takes value one in the column k if the user m is allocated to the carrier k and zero otherwise. Finally, replacing (6.19) in (6.18) allows to write the received signal as

$$\mathbf{y} = \boldsymbol{\Pi}\mathbf{s} + \mathbf{w}, \tag{6.20}$$

where $\boldsymbol{\Pi} = \sum_{m=0}^{L-1} \boldsymbol{\Pi}^{(m)}\boldsymbol{\Psi}^{(m)}$ is the overall interference matrix and $\boldsymbol{\Pi}^{(m)}$ is defined for each user as in (6.5). Note that $\boldsymbol{\Pi}$ in this section considers the interference of all users.

To compensate for CFO effects in the received multiuser block, it is necessary to obtain an estimate of $\mathbf{s}$, denoted $\hat{\mathbf{s}}$, from (6.20). Two estimation criteria can be considered: least squares (LS) and minimum mean squared error (MMSE) [32]. However, only LS is considered since for the SNR range of interest it has the same performance than MMSE with less complexity. LS compensation is given by

$$\hat{\mathbf{s}} = (\boldsymbol{\Pi}^H\boldsymbol{\Pi})^{-1}\boldsymbol{\Pi}^H\mathbf{y} = \boldsymbol{\Pi}^{-1}\mathbf{y}, \tag{6.21}$$

where the right-hand side of this equation is valid only if $|\xi^{(m)}| < 0.5$, i.e., if $\boldsymbol{\Pi}$ is a full rank, square matrix [32]. From (6.21) is noted that LS compensation requires the inversion of the interference matrix $\boldsymbol{\Pi}$, resulting in a high computational burden. These methods are denoted as *complete* compensation.

Given that the larger interference terms in $\boldsymbol{\Pi}$ are located around the main diagonal and taking into account the block circular symmetry of the matrix, a circular banded approximation of the interference matrix can be proposed [4]:

$$[\boldsymbol{\Pi}_{CB}]_{p,q} = \begin{cases} [\boldsymbol{\Pi}]_{p,q} & \text{if } |q - p| \leq \tau \\ [\boldsymbol{\Pi}]_{p,q} & \text{if } N - \tau \leq |q - p| \leq M - 1 \\ & \quad \& \quad \tau \geq 2N_{vs} + 1 \\ 0 & \text{otherwise.} \end{cases} \tag{6.22}$$

The matrix bandwidth τ controls the compensation capability and $\tau_2 = \tau - 2N_{vs}$. From (6.22) is noted that $M - 1 - (N - \tau) \geq 0$, which means $\tau \geq 2N_{vs} + 1$, since $N = M + 2N_{vs}$. As τ increases, the interference is further reduced at expense of increasing the complexity. This approximation leads to a significant complexity reduction in the calculation of $\hat{\mathbf{s}}$.

The substitution of (6.22) in (6.21) leads to the circular banded compensation (CBC) method, which requires the inversion of $\boldsymbol{\Pi}_{CB}$. By exploiting the banded matrix structure and the fact that most of the matrix components are zero, it is possible to derive low-complexity algorithms for the inversion of $\boldsymbol{\Pi}_{CB}$. These efficient algorithms can use, for example, LU decomposition and backward and forward substitution [33, 47]. The approximation in (6.22) allows a trade-off between complexity and interference cancellation. For the case when $\tau \leq 2N_{vs}$, the terms outside the main diagonal are masked by the virtual subcarriers. Then, the circular banded compensation coincides with the banded compensation (BC) proposed in [32].

The successive CFO compensation criterion takes advantage of the structure of the interference matrix $\mathbf{\Pi}$ defined in (6.4) [39]. The proposal is a formulation of the SAGE algorithm to compensate for the CFO, where the interference is suppressed iteratively by addressing the user's induced interference sequentially in each iteration. Considering the user m, the interference is suppressed in two stages: (1) the multiple access interference (MAI) (i.e., the interference of users $\{0 \ldots m - 1\, m + 1 \ldots L - 1\}$ over user m) is canceled, and (2) self-interference of user m is compensated.

The first step is based on the fact that the contribution of the signal of each user is added at the BS. As $\mathbf{s}^{(l)}$ for $l \in \{0, \ldots, m - 1, m + 1, \ldots, L - 1\}$ are not available at the receiver, they are replaced by previous estimates that will be updated iteratively as in the following expression:

$$\hat{\mathbf{y}}_i^{(m)} = \mathbf{y} - \mathbf{\Pi} \left(\sum_{l=0}^{m-1} \hat{\mathbf{s}}_{i+1}^{(l)} + \sum_{l=m+1}^{L-1} \hat{\mathbf{s}}_i^{(l)} \right), \tag{6.23}$$

where $\hat{\mathbf{y}}_i^{(m)}$ is an estimate of the received signal from user m at iteration i, $\mathbf{y}$ is the complete received signal, $\mathbf{\Pi}$ is the overall interference matrix, and $\hat{\mathbf{s}}_i^{(l)}$ are the symbol estimates of user l at iteration i. Note that the symbols of users $\{0, \ldots, m - 1\}$ are already updated, whereas those of users $\{m + 1, \ldots, L - 1\}$ correspond to the previous iteration.

In the second step, the MAI is assumed already compensated. Following LS for user m and considering $\mathbf{\Psi}^{(m)}$, the update equation results

$$\hat{\mathbf{s}}_{i+1}^{(m)} = \mathbf{\Psi}^{(m)} (\mathbf{\Pi}^{(m)})^{-1} \hat{\mathbf{y}}_i^{(m)}. \tag{6.24}$$

LS criterion in (6.24) coincides with the maximum likelihood estimation, since the noise is Gaussian.

If (I)DFTs in the algorithm described in the previous section are calculated using the (I)FFT algorithm, a great reduction in complexity is obtained compared to other iterative proposals. Therefore, this successive interference cancellation proposal is referred to FFT-based interference cancellation (FFT-IC). An upper bound for the mean square symbol error at convergence for the compensation algorithm is proposed in [48].

To compare the CFO compensation capabilities, Fig. 6.8 shows the bit error rate (BER) curves for FFT-IC [39], Direct [32], and conjugate gradient (CG) [37] algorithms. A 3GPP LTE-like system is considered, with 1024 subcarriers, a cyclic prefix of 152, and 16-QAM modulation. The system has 4 users, a tile size of 12 subcarriers, and 20 tiles per user with GCAS. The channel is considered known at the receiver and the symbols are recovered using a single-tap frequency domain

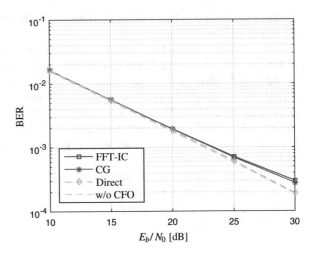

Fig. 6.8 BER performance comparison between different CFO compensation methods for OFDMA, considering GCAS for LS. The curves are obtained using the ITU Vehicular A channel and a CFO uniformly distributed in the interval $|\xi^{(m)}| < 0.3$

equalizer. The CFO range is $|\xi^{(m)}| < 0.3$. From the figure, it is clear that FFT-IC attains a similar BER than CG, but with a lower implementation complexity [39].

6.3.2   CFO Compensation for Multiuser FBMC

The FBMC transmission technique can be regarded as an extension of the OFDM concept. The OFDM rectangular time window is replaced by a highly selective filter that, in a multiple access context, leads to a negligible MAI [40]. As a consequence, only intercarrier interference needs to be compensated in multiuser FBMC.

The received signal can be described by

$$y(n) = \sum_{m=1}^{L} \left(\sum_{q=0}^{N_h-1} h_q^{(m)}(n) g^{(m)}(n-q) \right) e^{j2\pi \xi^{(m)} n/N} + w(n), \tag{6.25}$$

where $g^{(m)}(n)$ are symbols of the m-th user modulated by the synthesis filter bank [49], and $w(n)$ is the additive white Gaussian noise of variance σ^2. It is assumed that the timing error is absorbed by the channel model and it is compensated by the channel equalizer [24]. Since no CP is required in FBMC, the multicarrier symbol length is N.

For the analysis filter bank, illustrated in Fig. 6.9, a prototype filter with impulse response $p(n)$ and length $L_{ov}N - 1$ is used, where L_{ov} is the overlapping factor, i.e., the number of FBMC symbols equivalent to the prototype filter length. Also, $\beta_{m,v} = (-1)^{m(v+L_{ov})}$; $\phi_{m,v} = \{1, j, 1, j, \ldots\}$ for m even or $\phi_{m,v} = \{j, 1, j, 1, \ldots\}$ for m odd; and $b_m(n) = p(N-1-m+nN)$ is the m-th type-2 polyphase component

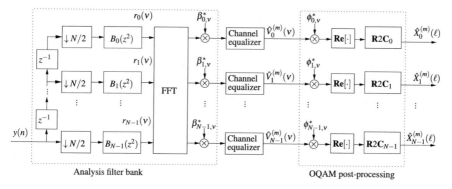

Fig. 6.9 Block diagram of filter bank multicarrier multiple access demodulation that consists of an analysis filter bank, a channel equalization, and a QAM/OQAM transformation

of $p(n)$ with z-transform $B(z^2)$. Finally, the output vector corresponding to the N branches of the polyphase decomposition at time ν can be described by

$$\mathbf{r}(\nu) = \sum_{q=0}^{L_{ov}-1} \mathbf{r}^{(q)}(\nu) = [r_0(\nu), \ldots, r_{N-1}(\nu)]^T, \qquad (6.26)$$

where

$$\mathbf{r}^{(q)}(\nu) = \mathbf{y}(\nu + 2q) \odot \mathbf{p}(L_{ov} - q),$$
$$\mathbf{y}(\nu) = [y(\nu N/2), \ldots, y(\nu N/2 + N - 1)]^T,$$
$$\mathbf{p}(l) = [p(lN), \ldots, p(lN - N + 1)]^T \qquad (6.27)$$

and $\odot$ denotes Hadamard vector product. Considering a multitap equalizer of L_{eq} coefficients, the symbols before the *offset-QAM* (OQAM) post-processing block are given by

$$\hat{\mathbf{V}}^{(m)}(\nu) = \sum_{l=0}^{L_{eq}-1} \mathbf{G}(\nu - l, \nu) \odot \boldsymbol{\beta}^*(\nu - l) \odot \mathscr{F}\left(\sum_{q=0}^{L_{ov}-1} \mathbf{r}^{(q)}(\nu - l)\right), \qquad (6.28)$$

where $\mathscr{F}\{\cdot\}$ is the FFT operator, $\hat{\mathbf{V}}^{(m)}(\nu) = [\hat{V}_0^{(m)}(\nu) \ldots \hat{V}_{N-1}^{(m)}(\nu)]^T$, $\hat{V}_k^{(m)}(\nu)$ is the output of the channel equalizer, $\boldsymbol{\beta}^*(\nu) = [\beta_{0,\nu}^* \ldots \beta_{N-1,\nu}^*]^T$, $\mathbf{G}(l, \nu) = [\mathscr{E}_0(l, \nu), \ldots, \mathscr{E}_{N-1}(l, \nu)]^T$, and $\mathscr{E}_k(l, \nu)$ is the multitap equalizer impulse response corresponding to time ν, subcarrier k (assigned to user m), for an impulse applied at l [24, Section 4.1]. Note that due to the double sampling frequency of the filter bank, the sampling rate of the channel is also doubled. After CFO compensation and channel equalization, the received OQAM symbols are converted into QAM symbols, as it is illustrated in Fig. 6.9.

The compensation is equivalent to the post-demodulation compensation for OFDMA, derived in [27], although it performs better due to the lower MAI inherent to multiuser FBMC.

The FBMC prototype filter $p(n)$ is designed to have low sidelobes to limit the intercarrier interference to a few subcarriers. For the special case of SCAS no multiple-access interference cancellation is required if few guard subcarriers are inserted between users. Therefore, the only intercarrier interference is the self-interference of each user. The direct cancellation method proposed in [27] can be used to eliminate the intercarrier interference of each user in the FBMC scheme of [41]. The intercarrier interference compensation is performed by counter-rotating the input to the receiver at the angular frequency given by the user's estimated CFO. That is, by replacing the input signal $y(n)$ by $y(n)e^{-j2\pi\hat{\xi}^{(m)}n/N}$ in (6.27), the compensation of $\mathbf{r}^{(q)}(v)$ is obtained as

$$\bar{\mathbf{r}}^{(q)}(v) = e^{-j\pi\hat{\xi}^{(m)}v} e^{-j2\pi\hat{\xi}^{(m)}q} \mathbf{c}(\hat{\xi}^{(m)}) \odot \mathbf{r}^{(q)}(v), \tag{6.29}$$

where $\mathbf{c}(\hat{\xi}^{(m)}) = [1 \quad e^{-j2\pi\hat{\xi}^{(m)}/N} \ldots e^{-j2\pi\hat{\xi}^{(m)}(N-1)/N}]^T$ and $\hat{\xi}^{(m)}$ denotes the CFO estimate of user m.

By inserting $\bar{\mathbf{r}}^{(q)}(v)$ in (6.28), it is obtained

$$\hat{\mathbf{V}}^{(m)}(v) = e^{-j\pi\hat{\xi}^{(m)}v} \left(\sum_{l=0}^{L_{eq}-1} \mathbf{G}(v-l, v)e^{j\pi\hat{\xi}^{(m)}l} \odot \boldsymbol{\beta}^*(v-l) \odot \right.$$

$$\left. \mathscr{F}\left\{ \sum_{q=0}^{L_{ov}-1} e^{-j2\pi\hat{\xi}^{(m)}q} \mathbf{c}(\hat{\xi}^{(m)}) \odot \mathbf{r}^{(q)}(v-l) \right\} \right). \tag{6.30}$$

Note that the CFO introduces the phase term $e^{j\pi\hat{\xi}^{(m)}l}$ that must be taken into account in the structure of the multitap equalizer shown in (6.30).

Equation (6.30) requires a large amount of calculations to remove the CFO interference of every user. Employing the replication identity [4] and defining

$$\mathbf{c}'(\hat{\xi}^{(m)}) = [\mathbf{c}^T(\hat{\xi}^{(m)}), e^{-j2\pi\hat{\xi}^{(m)}} \mathbf{c}^T(\hat{\xi}^{(m)}), \ldots, e^{-j2\pi\hat{\xi}^{(m)}(L_{ov}-1)} \mathbf{c}^T(\hat{\xi}^{(m)})]^T, \text{ and}$$

$$\mathbf{r}'(v) = [(\mathbf{r}^{(0)})^T, \ldots, (\mathbf{r}^{(L_{ov}-1)})^T]^T,$$

the compensated OQAM symbols can be written as:

$$\hat{\mathbf{V}}^{(m)}(v) = e^{-j\pi\hat{\xi}^{(m)}v} \left(\sum_{l=0}^{L_{eq}-1} \mathbf{G}(v-l, v)e^{j\pi\hat{\xi}^{(m)}l} \odot \boldsymbol{\beta}^*(v-l) \odot \right.$$

$$\left. \left[\mathscr{F}\left\{ \mathbf{c}'(\hat{\xi}^{(m)}) \right\} \circledast \mathscr{F}\left\{ \mathbf{r}'(v-l) \right\} \right|_{\downarrow L_{ov}} \right), \tag{6.31}$$

where ⊛ is the circular convolution and the FFTs have length $L_{ov}N$. As a small number of components of $\mathscr{F}\left\{\mathbf{c}'(\hat{\xi}^{(m)})\right\}$ have significant values, to reduce the complexity only the largest L_{cfo} are considered and the others are set to zero [40]. For one-tap equalization, the following compensation of the OQAM symbols results

$$\hat{\mathbf{V}}^{(m)}(v) = e^{-j\pi\hat{\xi}^{(m)}v}\mathbf{G}(v) \odot \boldsymbol{\beta}^*(v) \odot \left\lfloor \mathscr{F}\left\{\mathbf{c}'(\hat{\xi}^{(m)})\right\} \circledast \mathscr{F}\left\{\mathbf{r}'(v)\right\}\right\rfloor_{\downarrow L_{ov}}, \quad (6.32)$$

where $\mathbf{G}(v) = [H_0^{-1}(v), \ldots, H_{N-1}^{-1}(v)]^{\mathrm{T}}$ and $H_k(v)$ is the channel frequency response corresponding to user m, subcarrier k, and time v, if k belongs to m. A similar CFO compensation method is proposed in [40] for one-tap equalization and a different FBMC realization [50]. Note that the CFO-induced phase term in the multitap equalizer of (6.30) does not appear in the one-tap equalizer of (6.32).

An additional comment related to multiuser FBMC BER performance is important at this point. Different to the case of OFDMA, ICAS or GCAS cannot be used in multiuser FBMC since affordable solutions of low complexity are not available. Nevertheless, multiuser FBMC reduces the MAI which leads to a low performance degradation due to CFO.

If the CFO update is not required for every symbol, the resulting complexity reduces since it is not necessary to update the LU decomposition of matrix $\mathbf{\Pi}_{CB}$ in (6.21), and recalculate the vector $\mathscr{F}\left\{\mathbf{c}'(\hat{\xi}^{(m)})\right\}$ in (6.32).

The section is ended by comparing the complexity of OFDMA and multiuser FBMC compensation techniques. Additionally, the BER for low and high CFO is evaluated. Finally, the performance of the algorithm in a time varying CFO scenario (high mobility) is tested. Systems parameters are the same than in Fig. 6.8. For SCAS 205 carriers per channel are used. Only the banded approximation is considered for CFO compensation in OFDMA since a low-complexity technique is required for the comparison. The basic parameters for FBMC are $L_{ov} = 4$, one guard carrier, and 205 subcarriers per user. The prototype filter is designed using the frequency sampling technique [49] and a multitap equalizer with $L_{eq} = 3$ is adopted [24, Section 4.1].

The main complexity sources in the implementation are the NL_{ov}-size FFTs and the CFO compensation in FBMC, calculated for each received symbol, and the inversion of the CFO interference matrix in OFDMA, required for every new CFO estimation. As a direct consequence, if the CFO is required to be re-estimated too often the overall complexity of the OFDMA compensation results very high. In the following, the complexity of complete OFDMA and multiuser FBMC receivers is compared for three different CFO update rates: Case 1, *Time-invariant CFO*: With a constant CFO the update is not required, giving the lower bound in terms of implementation complexity; Case 2, *Slowly time-variant CFO*: The CFO is updated every N_u multicarrier symbols. Then, the multiplications involved in CFO update are divided by the update period N_u; and Case 3, *Highly time-variant CFO*: The CFO is updated every symbol, and it can be considered as the worst case of the OFDMA implementation complexity.

Fig. 6.10 Computational complexity of multicarrier receivers versus interference bandwidth τ for three cases of CFO: time invariant (case 1), slowly time-variant CFO (case 2), and highly time-variant (case 3)

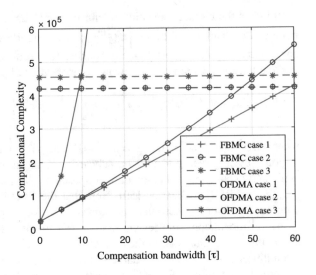

Fig. 6.11 OFDMA and multiuser FBMC CFO compensation schemes for fixed CFO considering the Vehicular A channel. In Fig. 6.11 is considered low CFO ($|\xi^{(m)}| < 0.1$) whereas in Fig. 6.12 high CFO ($|\xi^{(m)}| < 0.5$)

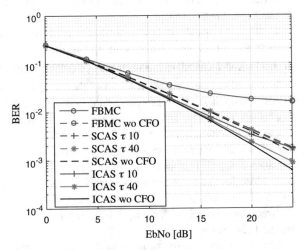

Figure 6.10 illustrates how the computational complexity of both systems varies as a function of the compensation bandwidth τ for three cases outlined above. $N_u = 200$ is considered for Case 2. For Case 1, OFDMA has lower complexity than multiuser FBMC for $\tau < 55$. For Case 2, $\tau < 45$ is required to have an OFDMA system with the lowest complexity. Finally, for Case 3, OFDMA is the less complex option if $\tau < 9$.

Figures 6.11 and 6.12 illustrate the BER performance of OFDMA and multiuser FBMC CFO compensation schemes, for the Vehicular A propagation channel [51] with a speed of 100 km/h. An invariant CFO uniformly distributed is considered. ICAS and SCAS are included for OFDMA, whereas multiuser FBMC is constrained to have SCAS. In Fig. 6.11, low CFO values ($|\xi^{(m)}| < 0.1$) are considered, whereas

Fig. 6.12 OFDMA and multiuser FBMC CFO compensation schemes for fixed CFO considering the Vehicular A channel. In Fig. 6.11 is considered low CFO ($|\xi^{(m)}| < 0.1$), whereas in Fig. 6.12 high CFO ($|\xi^{(m)}| < 0.5$)

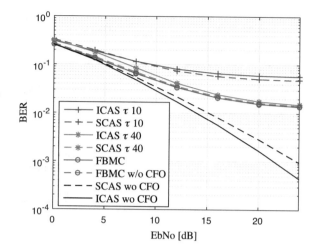

in Fig. 6.12 high values ($|\xi^{(m)}| < 0.5$) are used. From Fig. 6.11, it is noted that for low CFO, OFDMA outperforms FBMC both in BER and complexity regardless of the chosen CAS (see Fig. 6.10). On the other hand, Fig. 6.12 shows that multiuser FBMC has better performance than OFDMA for high CFO, except for $\tau = 40$ where the performance is similar. Using the compensation technique proposed for OFDMA, ICAS introduces frequency diversity but at the same time increases the MAI. As a consequence, for low CFO, ICAS has better performance than SCAS since the diversity gain prevails over the introduced MAI. On the other hand, for high CFO, SCAS is the best option due to the notorious increase in the MAI. Multiuser FBMC is not affected by the increase of CFO value since it has no MAI. Therefore, it admits higher carrier frequencies and lower quality local oscillators than OFDMA. The BER floor observed in multiuser FBMC is a consequence of the lack of robustness against channel variations. Since FBMC does not employ a cyclic prefix, it suffers from intersymbol interference (ISI) if the channel is time variant. FBMC equalization reduces the intercarrier (ICI) interference, but not the intersymbol interference. Then, for time-varying channels, the performance of the FBMC system is degraded.

Figure 6.13 illustrates the performance of time-varying CFO compensation in OFDMA and multiuser FBMC systems, for different CFO update periods N_u and subband CAS. Following the Rician-fading model in [3], the time-varying CFO is obtained from (6.1) considering $K_r = 15$, $v = 100$ km/h, and $d_m = 50$ m. The channel has an exponential decay profile given by $\sigma_l^2 = \exp(-9.2lT_s/1 \ \mu s)$ with $0 \leq l \leq N_h - 1$ and $N_h = 5$. From Fig. 6.13, it results that OFDMA is more robust to long CFO re-estimation periods, i.e., to higher N_u values. Despite variations in the CFO are considerable due to the high mobility assumption, long update periods are admissible (e.g., $N_u = 400$). As a consequence, the solution presented in Case 2 leads to a considerable complexity reduction. A recent test bench for high-speed trains was proposed in [52, Annex B], where $v = 350$ km/h and $d_m = 50$ m.

Fig. 6.13 OFDMA and
multiuser FBMC CFO
compensation schemes for
time-variant CFO considering
the Rician channel in [3]. The
CFO value is updated every
$N_u = 1$, $N_u = 400$, and
$N_u = 800$ multicarrier
symbols

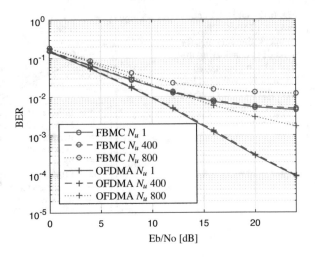

Considering $f_c = 2.5\,\text{GHz}$, $\Delta f = 15\,\text{kHz}$ [53], and a Doppler drift of 1% of Δf, the CFO needs to be re-estimated every $N_u \approx 1140$ multicarrier symbols or 190 subframes. This agrees with the results obtained in Fig. 6.13.

6.4   Summary of the Key Points

In this chapter, the effects of the carrier frequency offset (CFO) and the phase noise (PN) are analyzed. The CFO is produced by the difference between the frequency of the transmitter and the receiver oscillators, and the Doppler effect. On the other hand, the PN is caused by random fluctuations in the oscillators. Although the physical sources of CFO and PN are different, they can be included in a single OFDM frequency domain model (6.4). The interference has two components: a multiplicative common phase error (CPE) term and additive intercarrier interference (ICI) term.

For a user with low mobility, the CFO is almost constant and its estimation is required once, or at most a few times during transmission. However, if the user has high mobility, CFO estimation can be required several times during communication. Given the random nature of the PN, the estimation has to be done for each received symbol. PN estimation for moderate levels (Fig. 6.3a) is not complex, since the simple CPE compensation has good performance. For extreme cases (Fig. 6.3b), the ICI part also needs to be compensated, increasing the computational complexity.

Finally, the most relevant CFO and PN estimation and compensation techniques for single and multiuser cases are described. The best performance for CFO estimation in the single user case is achieved by [18]. CFO compensation techniques for OFDMA present a trade-off between performance and a computational complexity. The most relevant, ordered from the more simple to the most complex, are the least squares (LS) circular banded approximation [33], the FFT-based iterative cancellation [39], and the LS compensation [32]. Then, OFDMA and multiuser FBMC implementation complexity are analyzed, considering time varying CFO. It is shown that OFDMA performs better for low CFO since it takes advantage of the frequency diversity gain. On the other hand, multiuser FBMC performs better for high CFO since it is more robust to multiple access interference (MAI).

References

1. V. Syrjälä, M. Valkama, Analysis and mitigation of phase noise and sampling jitter in OFDM radio receivers. Int. J. Microw. Wirel. Technol. **2**(2), 193–202 (2010)
2. F. Horlin, A. Bourdoux, *Digital Compensation for Analog Front-Ends* (John Wiley & Sons, Hoboken, 2008)
3. S. Talbot, B. Farhang-Boroujeny, Time-varying carrier offsets in mobile OFDM. IEEE Trans. Commun. **57**, 2790–2798 (2009)
4. G. González, F. Gregorio, J. Cousseau, R. Wichman, S. Werner, Uplink CFO compensation for FBMC multiple access and OFDMA in a high mobility scenario. Phys. Commun. Elsevier **11**, 56–66 (2014)
5. D. Petrovic, W. Rave, G. Fettweis, Effects of phase noise on OFDM systems with and without PLL: characterization and compensation. IEEE Trans. Commun. **55**(8), 1607–1616 (2007)
6. T. Schenk, *RF Impairments in Multiple Antenna OFDM: Influence and Mitigation* (Technische Universitet Eindhoven, Eindhoven, 2006)
7. T.M. Schmidl, D.C. Cox, Robust frequency and timing synchronization for OFDM. IEEE Trans. Commun. **45**(12), 1613–1621 (1997)
8. P. Moose, A technique for orthogonal frequency division multiplexing frequency offset correction. IEEE Trans. Commun. **10**, 2908–2914 (1994)
9. M. Morelli, U. Mengali, Carrier-frequency estimation for transmissions over selective channels. IEEE Trans. Commun. **48**(9), 1580–1589 (2000)
10. B. Yang, K. Letaief, R. Cheng, Z. Cao, Channel estimation for OFDM transmission in multipath fading channels based on parametric channel modeling. IEEE Trans. Commun. **49**, 467–479 (2001)
11. F. Classen, H. Meyr, Frequency synchronization algorithms for OFDM systems suitable for communication over frequency selective fading channels, in *Proceedings of IEEE Vehicular Technology Conference (VTC)*, vol. 3 (1994), pp. 1655–1659
12. F. Daffara, O. Adami, A novel carrier recovery technique for orthogonal multicarrier systems. Eur. Trans. Telecommun. **7**(4), 323–334 (1996)
13. IEEE standard for local and metropolitan area networks part 16: Air interface for fixed and mobile broadband wireless access systems, amendment 2: Physical and medium access control layers for combined fixed and mobile operation in licensed bands and corrigendum 1. IEEE Std. 802.16 (2005)
14. 3rd Generation partnership project (3GPP) LTE; evolved universal terrestrial radio access (E-UTRA); physical layer procedures; (release 13), 3GPP TS 36.213 V13.2.0 (2016)
15. M. Ghogho, A. Swami, Unified framework for a class of frequency-offset estimation techniques for OFDM, in *IEEE International Conference on Acoustics, Speech, and Signal Processing (ICASSP)*, vol. 4 (2004), pp. iv–361–iv–364
16. M. Morelli, U. Mengali, An improved frequency offset estimator for OFDM applications. IEEE Commun. Lett. **3**(3), 75–77 (1999)
17. H. Minn, P. Tarasak, V.K. Bhargava, OFDM frequency offset estimation based on BLUE principle, in *IEEE Vehicular Technology Conference (VTC)*, vol. 2 (2002), pp. 1230–1234

18. G. González, F. Gregorio, J. Cousseau, S. Werner, R. Wichman, Data-aided CFO estimators based on the averaged cyclic autocorrelation. Signal Process. EURASIP **93**(1), 217–229 (2013)
19. V. Syrjälä, M. Valkama, N.N. Tchamov, J. Rinne, Phase noise modelling and mitigation techniques in OFDM communications systems, in *2009 Wireless Telecommunications Symposium* (2009), pp. 1–7
20. V. Syrjälä, M. Valkama, *Iterative Receiver Signal Processing for Joint Mitigation of Transmitter and Receiver Phase Noise in OFDM-Based Cognitive Radio Link* (IEEE, Piscataway, 2012)
21. P. Mathecken, S. Werner, T. Riihonen, R. Wichman, Subspace-based phase noise estimation in OFDM receivers, in *2015 IEEE International Conference on Acoustics, Speech and Signal Processing (ICASSP)* (2015), pp. 3227–3231
22. P.J. Mathecken, OFDM under Oscillator Phase Noise: Contributions to Analysis and Estimation Methods, Ph.D. Thesis, Department of Signal Processing and Acoustics, Aalto University (2016)
23. M.-O. Pun, M. Morelli, C.C. Jay Kuo, *Multi-Carrier Techniques For Broadband Wireless Communications: A Signal Processing Perspectives* (Imperial College Press, London, 2007)
24. PHYDYAS deliverable D3.1, Equalization and demodulation in the receiver (single antenna), in *PHYsical layer for DYnamic AccesS and cognitive radio* (2009)
25. Z. Cao, U. Tureli, Y.-D. Yao, Deterministic multiuser carrier-frequency offset estimation for interleaved OFDMA uplink. IEEE Trans. Commun. **52**(9), 1585–1594 (2004)
26. M.-O. Pun, M. Morelli, and C.-C.J. Kuo, Maximum-likelihood synchronization and channel estimation for OFDMA uplink transmissions. IEEE Trans. Commun. **54**(4), 726–736 (2006)
27. J. Choi, C. Lee, H.W. Jung, Y.H. Lee, Carrier frequency offset compensation for uplink of OFDM-FDMA systems. IEEE Commun. Lett. **4**(12), 414–416 (2000)
28. D. Huang, K.B. Letaief, An interference-cancellation scheme for carrier-frequency offsets correction in OFDMA systems. IEEE Trans. Commun. **53**(1), 203–204 (2005)
29. S. Manohar, D. Sreedhar, V. Tikiya, A. Chockalingam, Cancellation of multiuser interference due to carrier frequency offsets in uplink OFDMA. IEEE Trans. Wireless Commun. **6**(7), 2560–2571 (2007)
30. T. Yücek, H. Arslan, Carrier frequency offset compensation with successive cancellation in uplink OFDMA systems. IEEE Trans. Wireless Commun. **6**(10), 3546–3551 (2007)
31. P. Sun, L. Zhang, Low complexity iterative interference cancelation for OFDMA uplink with carrier frequency offsets, in *Proceedings of the 15th Asia-Pacific conference on Communications* (2009), pp. 390–393
32. Z. Cao, U. Tureli, Y.-D. Yao, Low-complexity orthogonal spectral signal construction for generalized OFDMA uplink with frequency synchronization errors. IEEE Trans. Veh. Technol. **56**(3), 1143–1154 (2007)
33. G. González, F. Gregorio, J. Cousseau, CFO compensation for OFDMA systems using circular banded matrices. Lat. Am. Appl. Res. (SPECIAL ISSUE – RPIC 2011) **43**(3), 255–260 (2014)
34. G. Chen, Y. Zhu, K.B. Letaief, Combined MMSE-FDE and interference cancellation for uplink SC-FDMA with carrier frequency offsets, in *Proceedings of IEEE International Conference on Communications* (2010), pp. 1–5
35. R. Fa, L. Zhang, R. Ramirez, A low-complexity grouped MMSE interference cancellation scheme for OFDMA uplink systems with carrier frequency offsets, in *Proceedings of IEEE Wireless Communications and Networking Conference* (2011), pp. 1602–1606
36. C.-Y. Hsu, W.-R. Wu, A low-complexity zero-forcing CFO compensation scheme for OFDMA uplink systems. IEEE Trans. Wireless Commun. **7**(10), 3657–3661 (2008)
37. K. Lee, S.-R. Lee, S.-H. Moon, I. Lee, MMSE-based CFO compensation for uplink OFDMA systems with conjugate gradient. IEEE Trans. Wireless Commun. **11**(8), 2767–2775 (2012)
38. J.A. Fessler, A.O. Hero, Space-alternating generalized expectation-maximization algorithm. IEEE Trans. Signal Process. **42**(10), 2664–2677 (1994)
39. L. Bai, Q. Yin, Frequency synchronization for the OFDMA uplink based on the tile structure of IEEE 802.16e. IEEE Trans. Veh. Technol. **61**(5), 2348–2353 (2012)

40. H. Saeedi-Sourck, Y. Wu, J.W.M. Bergmans, S. Sadri, B. Farhang-Boroujeny, Complexity and performance comparison of filter bank multicarrier and OFDM in uplink of multicarrier multiple access networks. IEEE Trans. Signal Process. **59**(4), 1907–1912 (2011)

41. P. Siohan, C. Siclet, N. Lacaille, Analysis and design of OFDM/OQAM systems based on filterbank theory. IEEE Trans. Signal Process. **50**(5), 1170–1183 (2002)

42. H. Holma, A. Toskala, *LTE for UMTS – OFDMA and SC-FDMA Based Radio Access*, 1st edn. (Wiley, Hoboken, 2009)

43. 3rd Generation partnership project (3GPP) technical specification group radio access network; physical layer aspects for evolved universal terrestrial radio access (UTRA) (release 7), 3GPP TR 25.814, V7.1.0 (2006)

44. E. Dahlman, S. Parkvall, J. Sköld, P. Beming, *3G Evolution HSPA and LTE for Mobile Broadband*, 2st edn. (Elsevier, Amsterdam, 2008)

45. IEEE Standard for local and metropolitan area networks part 16: air interface for fixed broadband wireless access systems, IEEE Std 802.16-2004 (Revision of IEEE Std 802.16-2001) (2004), pp. 01–857

46. IEEE Standard for local and metropolitan area networks part 16: air interface for broadband wireless access systems, amendment 3: Advanced air interface, IEEE Std. 802.16m (2011)

47. G.H. Golub, C.F. Van Loan, *Matrix Computations*, 3rd ed. (The Johns Hopkins University Press, Baltimore, 1996)

48. G. González, F. Gregorio, J. Cousseau, C. Muravchik, Analysis of the CFO successive interference cancellation for the OFDMA uplink. Springer Wirel. Pers. Commun. **91**(2), 989–1002 (2016)

49. PHYDYAS deliverable D5.1: Prototype filter and structure optimization, *PHYsical layer for DYnamic AccesS and Cognitive Radio* (2008)

50. N.J. Fliege, Computational efficiency of modified DFT polyphase filter banks, in *Proceedings of 27th Asilomar Conference on Signals, Systems and Computers*, vol. 2 (1993), pp. 1296–1300

51. ITU-R, Guidelines for evaluation of radio transmission technologies for IMT-2000. Recomm. ITUR M1225 **93**(3), 148–56 (1997)

52. LTE; evolved universal terrestrial radio access (E-UTRA); base station (BS) radio transmission and reception (2012)

53. 3rd Generation partnership project (3GPP) technical specification group radio access network; physical layer aspects for evolved universal terrestrial radio access (UTRA) (release 7) (2006)

Part III
RF Imperfection in Novel Technologies

This final part of the book starts with Chap. 7, where full-duplex (FD) transceiver design aspects are introduced. Feasible FD design requires different techniques to be used together. Main emphasis here is put on baseband signal processing for self-interference reduction. An important contribution here is related to consider the effects the non-ideal behavior of the basic transceiver blocks produce in the desired performance. Chapter 8 introduces massive MIMO systems. DL and UL aspects are considered in addition to channel non-reciprocity, antenna coupling, channel estimation, and pilot contamination. Minimum RF requirements (mostly ADC resolution) are key to the front-end design in this case and are discussed in detail, including numerical evaluations. Also, power consumption design aspects are considered. This part of the book also includes Chap. 9, where machine-type communications are considered. More specifically, internet of things applications and challenges are discussed. Main design concepts and parameters for IoT design are introduced. Emphasis is made on licensed IoT standards, mainly narrowband IoT (NB-IoT). In addition, a study of RF impairment effects from LTE on NB-IoT is included. Finally, this part concludes with Chap. 10 where some 5G implementation challenges are discussed, such as combination of massive MIMO and FD techniques, millimeter wave wireless design, massive MIMO challenges, and non-orthogonal multiple access schemes.

Chapter 7
Full-Duplex Communication Systems

Abstract In contrast to conventional half-duplex (HD) transceivers, a full-duplex (FD) transceiver is able to transmit and receive simultaneously in the same frequency band. The main impediment for FD implementation is the self-interference (SI) caused by the strong coupling of the transmitted signal to the receiver chain. Self-interference is usually mitigated by a combination of antenna cancellation and RF cancelers, both in the analog domain, followed by a digital cancellation technique. After analogue suppression, the SI can be tens of decibels above the signal of interest. Therefore, high-resolution ADCs are required in order to avoid the signal of interest be buried in the quantization noise. Additionally, high-speed ADCs are required to handle the large bandwidth channels envisioned for the next generation of communication systems. Full-duplex systems are very sensitive to different system distortions associated with the analog front-end electronics, since these imperfections significantly limit the SI suppression capability. The main sources of impairments are the nonlinear power amplifiers, phase noise from local oscillators, ADC quantization noise and distortion, and mismatches between I and Q branches of the transceivers. In this chapter, we first introduce a baseband interference model of the imperfections of the system components. Then, we introduce SI cancellation techniques, taking into account phase noise effects, mixer imbalances, and the power amplifier operation point. After that, we determine the necessary ADC requirements for the FD operation. Finally, we analyze the system performance in terms of the energy efficiency and spectral efficiency, considering different system parameters.

7.1 System Model

Last advances in electronics, integration techniques, and signal processing enabled real co-channel full-duplex (FD) operation in wireless transceivers, which is one of the most sought objectives since the invention of radio transmissions [1]. In contrast to conventional half-duplex (HD) transceivers, a FD transceiver is able to transmit and receive simultaneously in the same frequency band. Besides doubling

© Springer Nature Switzerland AG 2020

F. Gregorio et al., *Signal Processing Techniques for Power Efficient Wireless Communication Systems*, Signals and Communication Technology,
https://doi.org/10.1007/978-3-030-32437-7_7

the theoretical spectral efficiency, this may open up a wide range of new applications from cognitive radio spectral sensing [2] to more efficient medium access protocols [3–5].

7.1.1   Full-Duplex Transceivers

A block diagram of a FD transceiver is shown in Fig. 7.1, where the source (S) communicates with the receiver (R), and the transmitter (T) with the destination (D). The communication channels S-R and T-D are denoted as $h_{SR}(n)$ and $h_{TD}(n)$, respectively.

The main impediment for the FD implementation is the unavoidable self-interference (SI), caused by the coupling of the transmitted signal in the receiver chain. The SI takes into account not only the energy radiated by T that is captured by R, but also the internal interference produced at circuit level when the transmitter and receiver are implemented in the same board. For example, in FD transceivers of a single antenna, the isolation loss in the circulator contributes to the SI. In the figure, the SI channel is symbolized by the path between T and R, and denoted as $h_{SI}(n)$. This interference can be seen as a penalty for the FD operation, resulting in a spectral efficiency of less than two for real systems. Furthermore, for some harsh SI conditions, HD systems can outperform FD implementations [6].

In compact transceivers, the interference level is usually high compared to the signal of interest [7]. It depends on the transmitted signal power, the isolation between transmitter and receiver antennas, and the surrounding environmental reflectors. The SI is mitigated by a combination of passive and active antenna cancellation [8, 9] and RF cancelers [10, 11], both in the analog domain. In most applications, analog cancellation is not enough and a digital cancellation stage is also required [12–15].

Fig. 7.1 FD transceiver model

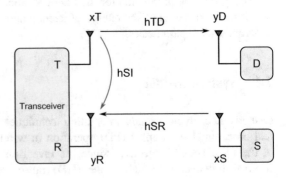

Following the signal flow in Fig. 7.1 and considering a baseband digital signal model, the received signals at the destination and the receiver can be expressed as:

$$y_R(n) = h_{SR}(n) * x_S(n) + h_{SI}(n) * x_T(n) + w_R(n). \tag{7.1}$$

$$y_D(n) = h_{TD}(n) * x_T(n) + w_D(n), \tag{7.2}$$

where $w_D(n)$ and $w_R(n)$ are AWGN terms with variances σ_D^2 and σ_R^2, respectively. As it is noted from (7.1), the performance of the receiver is compromised by the SI. In Sect. 7.2, some SI cancellation techniques are presented and their performance analyzed considering several impairments in the FD transceiver.

7.1.2 Full-Duplex Relays

The inclusion of relays in communication systems leads to an increase in the system coverage area and throughput, reduces the source transmitted power, and provides additional diversity and reliability. FD relays increment the spectral efficiency since only one channel is needed, in comparison to HD relays that need two channels (in time or frequency). An interesting application example is the high-speed train relayed link [16]. In this scenario, considering OFDM modulation, the intercarrier spacing is increased to make the relay-train link more robust against speed related effects. On the other hand, as the base station-relay link is static, the intercarrier spacing remains the same. The complex interference that results from the different modulation parameters is analyzed in [17].

A block diagram of a two-hop FD relaying link is illustrated in Fig. 7.2. In this case, the S-R (source-relay) link is the first hop, and T–D (relay–destination) is the second hop. In some applications, the direct link S–D could not exist due to, e.g., shadowing in urban deployments or large-scale fading in large coverage networks.

The most common operation modes for relays are amplify-and-forward (AF) and decode-and-forward (DF). In the AF mode, the relay terminal simply amplifies

Fig. 7.2 Two-hop relay link model

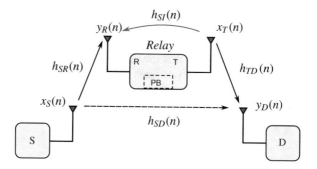

and retransmits the received signal from the source to the destination. On the other hand, in the DF mode, the received signal is first demodulated and decoded before retransmission. Although the AF method is simple, it suffers from noise amplification and stability problems, as will be shown later. On the other hand, DF has a nonlinear behavior. The relay is able to decode the received symbols and send an error-free message to the destination, performing better than AF. Alternatively, for low SINR it makes errors in the decisions that propagate to the destination causing a significant performance drop. It should be noted that for block-based transmissions, like OFDM, a minimum delay of one block is needed to make the decoding. There is another relaying strategy known as compress-and-forward (CF), where the relay sends to the destination a compressed version of the received message. In some circumstances, this strategy can perform better than AF and DF [18]. The relay operation mode is modeled in Fig. 7.2 as the processing block (PB).

For the DF operation mode and assuming error-free decoding, the processing block has input–output relation $x_T(n) = \kappa_R x_S(n-\tau)$, where κ_R is a scalar modeling the processing gain and τ the processing delay. Then, the received signals at the relay and destination are

$$y_R(n) = h_{SR}(n) * x_S(n) + \kappa_R h_{SI}(n) * x_S(n - \tau) + w_R(n) \qquad (7.3)$$

$$y_D(n) = \kappa_R h_{TD}(n) * x_S(n - \tau) + w_D(n). \qquad (7.4)$$

From (7.3) is clear that even considering decoding without errors, the SI reduces the SINR in the first hop and must be mitigated [19].

On the other hand, the processing block for an AF relay is expressed as $x_T(n) = \kappa_R y_R(n - \tau)$. The received signals for this case are

$$y_R(n) = \kappa_R h_{SR}(n) * x_S(n) + \kappa_R h_{SI}(n) * y_R(n - \tau) + w_R(n) \qquad (7.5)$$

$$y_D(n) = \kappa_R h_{TD}(n) * y_R(n - \tau) + w_D(n). \qquad (7.6)$$

Then, using the z-transform and replacing (7.5) in (7.6), the destination signal can be expressed as:

$$Y_D(z) = \frac{\kappa_R H_{TD}(z) H_{SR}(z) z^{-\tau} X_S(z)}{1 - \kappa_R z^{-\tau} H_{SI}(z)} + \frac{\kappa_R H_{TD}(z) W_R(z)}{1 - \kappa_R z^{-\tau} H_{SI}(z)} + W_D(z). \qquad (7.7)$$

Equation (7.7) reveals that the FD AF relay could oscillate or even be unstable (see first r.h.s. term) and it produces the amplification of the thermal noise at the relay input (second r.h.s. term). For this reason, the SI cancellation technique and the relay gain need to be adjusted properly to ensure a correct operation.

7.2   Self-interference Removal

As can be concluded from the previous section, the SI mitigation is the key enabling factor for the implementation of FD systems. This task is usually performed in three stages: antenna, RF, and digital cancellation, as illustrated in Fig. 7.3a.

Antenna cancellation is essential since it is the first stage in the interference reduction process. It aims to reduce the physical coupling between the transmitter and the receiver radiating elements. Passive techniques optimize the distance and the orientation of the antennas, and include absorbers to lower the interference. On the other hand, active techniques are based on the use of multiple antennas and beamforming. The best cancellation performance is reported in the literature when passive and active methods are combined together. An antenna system for SI cancellation in FD systems is proposed in [20], and shown in Fig. 7.4, where a two-element array is used for transmission and another for the reception. A great isolation between the transmitter and receiver ports is achieved, given the aggregate effect of the change in polarization between the arrays and the shaping of the transmitter radiation pattern.

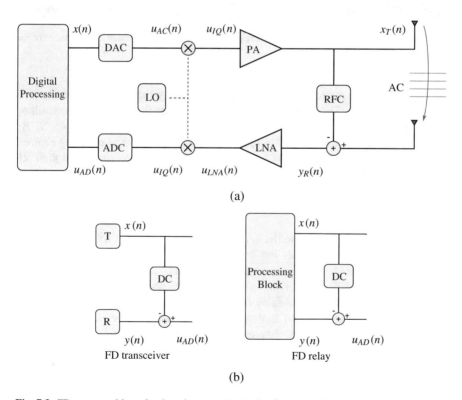

Fig. 7.3 FD system with analog impairments. (**a**) Analog front-end. (**b**) Digital processing block

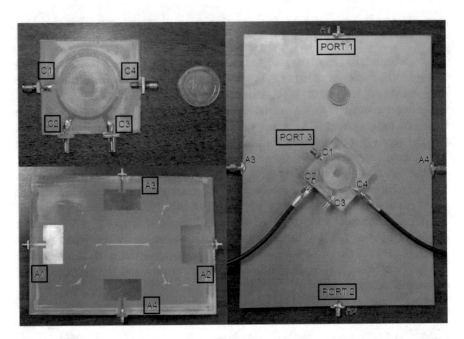

Fig. 7.4 FD patch antennas. A two-element array is used for transmission and another for the reception

The antenna SI reduction technique is followed by the RF cancellation, as is shown in Fig. 7.3a. This second stage subtracts a contribution of the transmitted signal at the receiver, aiming for a substantial reduction in the SI power. By defining, respectively, the digital baseband model of the physical coupling channel and the single tap RF canceler as $\tilde{h}_{SI}(n)$ and $h_{RF}(n)$, the received signal after RF cancellation can be expressed as:

$$y_R(n) = h_{SR}(n) * x_S(n) + (\tilde{h}_{SI}(n) - h_{RF}) * x_T(n) + w_R(n). \qquad (7.8)$$

Then, if the residual SI is defined as $h_{SI}(n) = \tilde{h}_{SI}(n) - h_{RF}(n)$, (7.8) becomes (7.1). This simplified approach does not consider any imperfection in the transceiver components. Those non-idealities are treated in Sects. 7.2.1–7.2.4. It must be noted that although the RF canceler has a single tap, its baseband model could have multiple taps depending on the delay and system bandwidth.

Since the RF canceler is an analog filter, it is usually restricted to be a simple attenuation and delay, due to implementation constraints. There are nowadays commercial off-the-shelf vector modulators designed for this task, as the one presented in [21]. These cancelers allow to digitally control both attenuation and delay, which make them suitable for on-line coefficient adaptation. Coefficients are tuned using an adaptive procedure to minimize the received SI power.

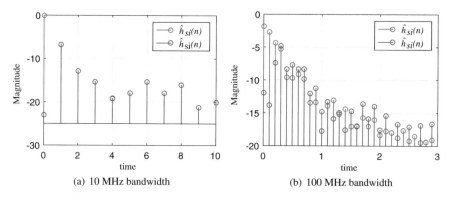

Fig. 7.5 Residual channel impulse response after RF cancellation stage: (**a**) narrowband channel, (**b**) broadband channel

To illustrate the result of the RF cancellation procedure, Fig. 7.5 shows the SI channel before and after the RF cancellation. The RF cancellation was applied on measurements of the SI channel taken using the FD antenna [20], for bandwidth of 10 and 100 MHz, and centered at 2.5 GHz. In Fig. 7.5a is noted that the RF canceler compensates almost exclusively for the first path, since the contribution of the reflecting paths is concentrated in this first coefficient due to the small bandwidth (time resolution is 1 μs). The total interference reduction is 12.87 dB. On the other hand, Fig. 7.5b shows that the RF canceler produces a significant reduction in the first three channel paths. In this case, the larger time resolution allows the distinction of more reflections (time resolution is 0.1 μs). In this case, the interference reduction is 5.78 dB. The notorious drop in the cancellation performance indicates that for broadband channels a multitap RF canceler may be needed. In both cases, the resulting digital channel has less interference power but it is more frequency selective.

Regarding the statistical model, the physical SI channel $\tilde{h}_{SI}(n)$ can be modeled as Rician due to the strong coupling between the transmitter and receiver antennas. This coupling is significant, because transmitter and receiver antennas are typically close to each other in a FD device, or transmitter and receiver chains may even share the same antenna array [8]. In addition to the specular path, independent coefficients are included to model reflections on surrounding objects. Since the RF cancellation mainly reduces the SI power related to the coupling path, as it concentrates most of the energy, the SI residual channel $h_{SI}(n)$ results more frequency selective. Therefore, it is more adequately modeled as Rayleigh, as verified by the results shown in Fig. 7.5

Analog stages reduce the SI in order to relax the specifications of the low noise amplifier and to avoid the desired signal to be drowned by the analog-to-digital converter (ADC) quantization noise [22]. A throughout analysis of the ADC requirements for FD systems is presented in Sect. 7.3.

Once in the digital domain, advanced processing techniques can be applied to mitigate the residual interference. For that purpose, the receiver obtains an estimate of the residual SI channel $\hat{h}_{SI}(n)$ using any of the classic estimation techniques based on known preambles or dedicated pilots. Then, assuming a receiver without imperfections $(r_{AD}(n) = y_R(n)$ in Fig. 7.3a, b), the received signal after digital cancellation can be written as:

$$
\begin{aligned}
y(n) &= h_{SR}(n) * x_S(n) + h_{SI}(n) * x_T(n) - \hat{h}_{SI} * x(n) + w_R(n) \\
&= h_{SR}(n) * x_S(n) + (h_{SI}(n) - \hat{h}_{SI}(n)) * x(n) + w_R(n),
\end{aligned}
\tag{7.9}
$$

where a transmitter without imperfections is assumed in the second line $(x(n) = x_T(n))$. The difference $h_{SI}(n) - \hat{h}_{SI}$ represents the residual interference after the three cancellation stages, and is the penalty for the FD operation mode.

After analog cancellation and considering ideal components, it is easy to show that the signal received at the relay results the same as (7.3) for a DF relay or (7.5) for an AF relay. When digital cancellation is introduced, the received signal for a DF relay results

$$
y_R(n) = h_{SR}(n) * x_S(n) + \kappa_R(h_{SI}(n) - \hat{h}_{SI}(n)) * x_S(n - \tau) + w_R(n). \tag{7.10}
$$

It is evident that if $\hat{h}_{SI}(n)$ is a good approximation of $h_{SI}(n)$, the digital cancellation will reduce considerably the amount of errors in the symbols decoding. On the other hand, for AF relays the received signal with digital cancellation becomes

$$
y_R(n) = \kappa_R h_{SR}(n) * x_S(n) + \kappa_R(h_{SI}(n) - \hat{h}_{SI}(n)) * y_R(n - \tau) + w_R(n). \tag{7.11}
$$

Here, it is easy to show that if digital cancellation is well designed, it could be able to reduce the harmful effects of the feedback loop analyzed in (7.7). Some examples of digital SI cancellation filters are presented in Sects. 7.2.2 and 7.2.3.

Some experimental FD transceivers with distortion are able to double the rate of a HD system at an SNR of 20 dB, which is the maximum theoretical gain [23]. However, these results are difficult to achieve outside the laboratory due to the noise effect and imperfections in the transceiver implementation introduced by off-the-shelf components. In other words, the cancellation in (7.9)–(7.11) is far for being perfect. For this reason, these impairments should be taken into account to ensure low-cost implementations of FD systems. Additionally, some of these imperfections can be jointly mitigated with SI cancellation. In the following section some of these solutions are presented.

7.2.1   Full-Duplex Transceivers with Hardware Impairments

Before presenting SI and hardware imperfections cancellation techniques, some discussion about the baseband discrete models is presented. The most harmful impairments for the performance of the FD system are the power amplifier nonlinearity, the I/Q imbalance, and the quantization noise in the ADC. The DAC and the LNA do not introduce significant distortion, and are thus considered ideal devices. These components are included in the block diagram of Fig. 7.1. The baseband models of the impairments are described as follows.

The imbalance in the upconversion mixer can be modeled as:

$$u_{IQ}(n) = [g_{1T}u_T(n) + g_{2T}u_T^*(n)]e^{jv_T(n)}, \qquad (7.12)$$

where g_{1T} and g_{2T} denote the coefficients associated with the amplitude and phase mismatches in the transmitter, defined in Sect. 2.7; and $v_T(n)$ is the phase noise, defined in Sect. 2.8. The output of the power amplifier (PA) is given by

$$x_T(n) = K_{PA}f_{PA}\{u_{IQ}(n)\}, \qquad (7.13)$$

where K_{PA} is the PA gain and $f_{PA}\{\cdot\}$ is the PA nonlinear response. Several models to describe the behavior of the PA are presented in Sect. 2.5.2.

In a similar way as that of the upconversion process, the signal at the output of the receiver mixer can be expressed as:

$$r_{IQ}(n) = [g_{1R}y_R(n) + g_{2R}y_R^*(n)]e^{jv_R(n)}, \qquad (7.14)$$

where g_{1R} and g_{2R} denote the coefficients associated with the amplitude and phase mismatches in the receiver, and $v_R(n)$ is the phase noise. For the sake of generality, it is assumed that the parameters of the transmitter and the receiver can be different. Finally, after digital conversion the signal results

$$r_{AD}(n) = f_{AD}\{r_{IQ}(n)\}, \qquad (7.15)$$

where $f_{AD}\{\cdot\}$ denotes the nonlinear ADC model that describes the quantization noise, defined in Sect. 5.1.

In the following sections, techniques for SI channel mitigation considering hardware impairments are presented. Additionally, the FD system performance is analyzed using different figures of merit (SNR loss, BER, in- and out-of-band distortions).

7.2.2 Nonlinear Power Amplifier

The most harmful impairment for FD systems is by far the PA nonlinearity. For this reason, the topic was addressed in many articles in the literature. In particular, the SI cancellation taking into account the effect of the power amplifier (PA) nonlinearity is considered in [3, 24–27]. Most of the proposed solutions focus on the in-band cancellation without considering the out-of-band distortion generated by the PA

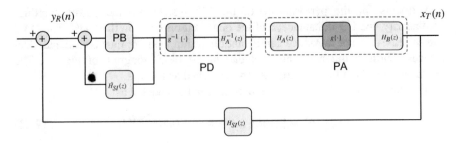

Fig. 7.6 Block diagram of full-duplex relay node with memory predistorter

nonlinearity. The resulting growth in out-of-band emission reduces the spectral efficiency of FD systems, because guard bands are needed to limit adjacent channel interference.

To model accurately the SI and the nonlinearity with memory effects, a Wiener–Hammerstein model can be used. This model, that consists of the cascade of a linear filter $H_A(n)$, a nonlinear static function $g\{\cdot\}$, and another linear filter $H_B(n)$, has excellent modeling capabilities for this case. Then, considering a FD relay, the SI canceler and predistorter (PD) that jointly compensate for the impairments follow the structure shown in Fig. 7.6 [28]. In contrast to other full-duplex PD designs [29], the solution presented here does not need an extra RF chain to estimate the PA response since it exploits the inherent SI signal. Although the proposed model is derived for a relay, the same principle can also be applied to a FD transceiver.

A two-step procedure is proposed to estimate the PD and SI canceler parameters. First, the linear part of the model is obtained using a low peak-to-average-power ratio (PAPR) training sequence. The low PAPR sequence is designed to avoid the nonlinear PA behavior. From Fig. 7.6, it is easy to note that the closed loop transfer function, without the SI canceler and the PD, is the cascade of $H_A(z)$, $H_B(z)$, and $H_{SI}(z)$.

Then, in order to estimate the nonlinear block, an OFDM sequence that excites the PA in its complete dynamic range is used. In the absence of noise and assuming the linear part of the model is perfectly equalized, it is possible to write

$$Y_{eq}(z) = g[X(z)H_A(z)]H_B(z)H_{SI}(z)H_L(z)$$
$$= g[X(z)H_A(z)]/H_A(z), \qquad (7.16)$$

where $H_L(z) = 1/(H_A(z)H_B(z)H_{SI}(z))$. This relation makes possible to estimate $H_A(z)$ and $g^{-1}\{\cdot\}$, by adjusting their coefficients to minimize the squared error $|e(n)|^2$, as defined in Fig. 7.7. More details of the error minimization are found in [28].

Fig. 7.7 FD transceiver model

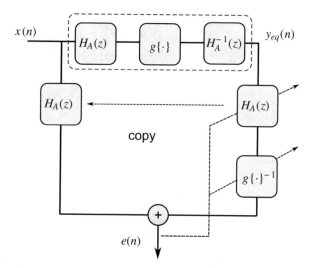

Finally, the coefficients of the equivalent SI channel (the cascade of $H_B(z)$ and $H_{SI}(z)$) and the parameters of the PD ($H_A(z)$ and the inverse of nonlinear block $g\{\cdot\}$) shown in Fig. 7.6 can be obtained from the procedure of two steps explained before.

The performance of the PD is evaluated in a FD relay system with OFDM modulation, 512 carriers, and a bandwidth of 10 MHz. The antenna and RF cancellation are equivalent to 50 and 30 dB of attenuation, respectively. The SNR at the relay input is assumed to be 30 dB. The SI channel is modeled as Rician with a factor K_r of 30 dB. At initialization, one OFDM symbol with 32 active subcarriers is used to identify the linear cascade, and 20 OFDM symbols for the nonlinear block. The predistorter polynomial has order $P = 7$ and $H_L(z)$ has 5 coefficients. The power amplifier is modeled as a Wiener–Hammerstein system, where the linear filters are given by $H_A(z) = \frac{1+0.1z^{-2}}{1-0.1z^{-1}}$ and $H_B(z) = \frac{1-0.1z^{-1}}{1-0.2z^{-1}}$ [30], and the static nonlinearity follows a solid state power amplifier (SSPA) model with smoothing factor $p = 2$.

Figure 7.8a illustrates the SINR at the relay output, obtained for the PA operating with different IBO. The figure compares the performance of the PD with a linear SI canceler, a SI nonlinear canceler (NLC), and a SI linear canceler (LC). The PD outperforms NLC and LC techniques in terms of SINR for severe PA distortion. On the other hand, the performance in terms of the out-of-band distortion is shown in Fig. 7.8b. The figure shows the power spectral density (PSD) at the relay output for the same compensation techniques. Again, the advantage of the PD technique is verified. Finally, Table 7.1 illustrates the adjacent channel leakage for the same cases used in Fig. 7.8a.

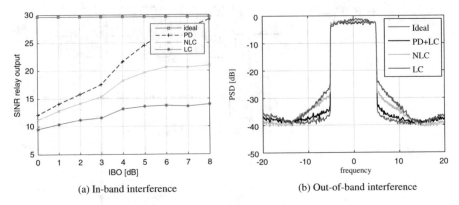

(a) In-band interference   (b) Out-of-band interference

Fig. 7.8 Predistorter performance. (**a**) In-band distortion when the PA operates with different IBOs. (**b**) Out-of-band distortion when PA operates with an IBO of 4 dB

Table 7.1 Adjacent channel power leakage ratio (ACLR) for different PA back-off levels; the values are calculated for an operation band of 10 MHz when the first adjacent band is at 7–12 MHz

PA IBO	LC	NLC	PD+LC
2 dB	−30.4 dB	−34.0 dB	−41.0 dB
4 dB	−33.5 dB	−38.4 dB	−45.9 dB
6 dB	−37.3 dB	−42.9 dB	−46.5 dB

7.2.3 I/Q Imbalance and Optimization of the Power Amplifier Operation Point

The I/Q imbalance in the transmitter and receiver mixers is also responsible for a large performance degradation if it is not properly compensated. The joint SI and I/Q cancellation is derived in [13] for a generic FD transceiver and in [31] for an AF FD relay. Given the improper nature of the I/Q imbalance, those proposals are based on widely linear filters. Interesting enough is that the filter structure makes the interference improper, after the SI and I/Q compensation.

Considering a FD relay as shown in Fig. 7.3b, the cancellation of the SI and source-relay (SR) channel can be obtained using the following widely linear filter [31]:

$$x(n) = h_1(n) * r_{AD}(n) + h_2(n) * r^*_{AD}(n), \tag{7.17}$$

where

$$h_1(n) = \frac{g^*_{1R} g^*_{1T} c(n) + g_{2T} g^*_{1R} c^*(n)}{(|g_{1R}|^2 - |g_{2R}|^2)(|g_{1T}|^2 - |g_{2T}|^2)} \tag{7.18}$$

$$h_2(n) = \frac{-g_{2R}g_{1T}^* c(n) - g_{1R}g_{2T}c^*(n)}{(|g_{1R}|^2 - |g_{2R}|^2)(|g_{1T}|^2 - |g_{2T}|^2)}. \tag{7.19}$$

The z-transform of the filter $c(n)$ that compensates for the SR and the SI channels is given by

$$C(z) = \frac{1}{K_{AD}[H_{SR}(z) + K_{PA}H_{SI}(z)z^{-\tau}]}, \tag{7.20}$$

where K_{AD} and K_{PA} are the compression term of the ADC and the PA gain, respectively. The derivation of the coefficients (7.18)–(7.20) can be found in [32]. It must be noted that the proposed canceler implements jointly the PB and digital canceler functions shown in Fig. 7.3b.

The relay digital output after I/Q, SI, and SR channel compensation can be written as:

$$x(n) = K_{PA}r_{AD}(n - \tau) + K_{PA}\Psi_1 w_{AD}(n) * c^*(n) + K_{PA}\Psi_2 w_{AD}(n)^* * c(n)$$
$$+ K_{PA}w(n) * c(n - \tau) + w_{PA}(n), \tag{7.21}$$

where

$$\Psi_1 = \frac{g_{1R}^* g_{1T} g_{2T} - g_{2R}^* g_{1T} g_{2T}}{(|g_{1R}|^2 - |g_{2R}|^2)(|g_{1T}|^2 - |g_{2T}|^2)} \tag{7.22}$$

$$\Psi_2 = \frac{|g_{2T}|^2 g_{1R} - |g_{1T}|^2 g_{2R}}{(|g_{1R}|^2 - |g_{2R}|^2)(|g_{1T}|^2 - |g_{2T}|^2)}. \tag{7.23}$$

It is easy to note than the inclusion of the compensation results in non-proper noise terms in (7.21).

The PA nonlinearity level depends on the PA IBO [3]. The higher the IBO, the lower the PA nonlinear effect but also the lower the PA efficiency. This means that the IBO sets a trade-off in the SINR of a two-hop relay link. If the IBO decreases, the SNR at the destination increases, but at the same time this also increases PA nonlinear interference level and the SI, reducing the SINR at the relay input. As a consequence, an optimal IBO that maximizes the SINR at destination can be found.

Assuming the DFT of $c(n)$ and $h_{TD}(n)$, respectively, denoted as $C(k)$ and $H_{TD}(k)$, has Gaussian distribution and disregarding the ADC impairment, the SINR at the destination of a two-hop link with OFDM modulation can be bounded as:

$$\frac{NP_x K_{PA}^2(\nu)}{\sigma_D^2 + 2\{K_{PA}^2(\nu)\sigma_R^2 + \sigma_{PA}^2(\nu)\}} < \text{SINR}_d \leq \frac{NP_x K_{PA}^2(\nu)}{\sigma_D^2 + \sigma_{PA}^2(\nu)}, \tag{7.24}$$

where N is the number of subcarriers, $w_{PA}(n)$ and $w_D(n)$ are considered normalized white processes, and the transmitter power P_x is equally distributed among the subcarriers. The dependency of K_{PA}^2 and σ_{PA}^2 on the IBO ν is included.

The IBO that maximizes the SINR at the destinations can be obtained by maximizing the lower and upper bounds in (7.24). For the lower bound, the IBO can be obtained from

$$\frac{dK_{PA}(v)}{dv}\left(\sigma_D^2 + 2\sigma_{PA}^2(v)\right) - K_{PA}(v)\frac{d\sigma_{PA}^2(v)}{dv} = 0. \qquad (7.25)$$

It is interesting to note that (7.25) does not depend on σ_R^2. This means that it is not possible to mitigate the effect of the noise at the relay input by varying the PA IBO. In a similar way, the IBO for the upper bound can be found from

$$2\frac{dK_{PA}(v)}{dv}(\sigma_D^2 + \sigma_{PA}^2(v)) - K_{PA}(v)\frac{d\sigma_{PA}^2(v)}{dv} = 0. \qquad (7.26)$$

For the soft limiter (SL) PA model there exists a closed expression for $K_{PA}(v)$ and $\sigma_{PA}^2(v)$, as defined in Sect. 2.5.2. Then, a closed-form solution to (7.25) and (7.26) is presented as follows. The result for the lower bound is

$$\frac{\text{erfc}(v)}{v} = \frac{\sigma_D^2}{A_c^2\sqrt{\pi}}, \qquad (7.27)$$

where $\text{erfc}(v)$ is the complementary error function and A_c the PA clipping level. Similarly, for the upper bound we have

$$\frac{\text{erfc}(v)}{v} = \frac{2\sigma_D^2}{A_c^2\sqrt{\pi}}. \qquad (7.28)$$

Finally, if we define the solutions of (7.27) and (7.28) as v_{LB}^2 and v_{UB}^2, respectively; the optimal IBO for the SL model is bounded by $v_{LB}^2 < v_{op}^2 \leq v_{UB}^2$. If the expressions of $K_{PA}(v)$, $P_{PA}(v)$, and their derivatives are not available, (7.26) and (7.25) can be solved numerically.

For the simulation, a FD relay with OFDM modulation, 1024 subcarriers, and a cyclic prefix of 64 samples is considered. A practical I/Q imbalance of $\beta_R = \beta_T = 1.02$, $\theta_R = \theta_T = 1.5°$, corresponding to an image rejection ratio (IRR) of 35 dB is added. We consider a SNR of 20 dB, an ADC of 12 bits, $K_{AD} = 1$, a PAPR of 13 dB, and a delay of $\tau = 5$. The PA parameters are $A_s = 1$ for SL and $A_s = 1$, $p = 2$ for SSPA. The SR channel is static 3 taps and the SI channel is a Rician channel with also 3 taps and a K_r-factor of 5 dB.

Figure 7.9 shows the result of the numerical SINR maximization and BER minimization v^2, and compares them with the derived theoretical bounds (7.27) and (7.28), considering $\text{SNR}_d = 10, 12, \ldots, 30$ dB. As expected, maximum values lie between the upper and lower bounds. It should be noted that the upper bound underestimates the optimum IBO, whereas the lower bound overestimates it. Even when the error in IBO for BER optimization goes from 0.55 dB for $\text{SNR}_d = 10$ dB

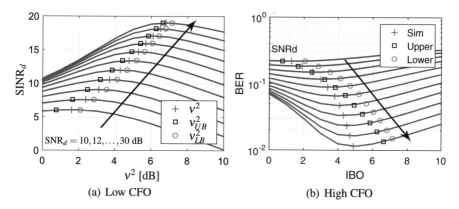

Fig. 7.9 Maximization of SINR (**a**) and minimization of BER (**b**) for a soft limiter model and $\text{SNR}_d = 10, 12, \ldots, 30\,\text{dB}$. Rayleigh relay-destination channel

to 1.73 dB for $\text{SNR}_d = 30$ dB, the BER difference between the optimal value and the theoretical bounds is not very large since BER curves are rather flat around the optimum.

7.2.4 Influence of Impairments in the System Performance

To conclude the section, the effect of the oscillator phase noise, the limited resolution of the ADC, the I/Q imbalance, and the power amplifier in a FD system is analyzed.

For the simulations, a FD relay scenario is evaluated assuming OFDM modulation with 512 subcarriers and a bandwidth of $B = 5$ MHz. Two levels of antenna cancellation are considered: $A_c = 30$ dB, corresponding to passive antenna cancellation, and 50 dB achieved with an active antenna cancellation technique. An RF cancellation $A_{rf} = 30$ dB is assumed. Three TX and RX I/Q imbalance scenarios are considered: ideal mixers, moderate imbalance with an image-rejection ratio (IRR) of 45 dB, and severe imbalance with $IRR = 35$ dB. The phase noise of the LO signal is specified by a 3 dB bandwidth of 100 Hz. The PA is modeled with a third-order polynomial with coefficients $1.1136 + j0.184$ and $-0.3807 - j0.0705$. A linear PA (LPA) is included for comparison purpose. The SI channel is modeled as Rician.

The difference between the SNR at relay input and the SINR at the output, namely $L = SNR_R - SINR_o$, is used to measure the influence of the impairments. This value is useful to evaluate the quality of the relay in terms of the used self-interference cancellation method and the robustness against the RF front-end imperfections. The SNR at relay input is $SNR_R = 30$ dB.

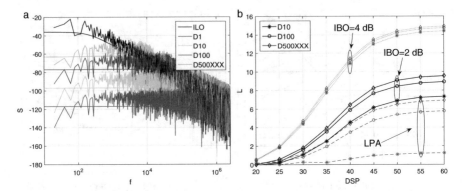

Fig. 7.10 Analysis of the local oscillator PN in a FD relay. (**a**) Power spectral density of receiver/transmitter cascade. $f_{3dB} = 100\,\text{Hz}$ and processing delay $\tau = 1, 10, 100, 500$ samples. Independent local oscillator (ILO) is included as reference and is equivalent of an infinite processing delay. (**b**) SNR loss curves: Nonlinear distortion and processing delay

The power spectral density of the local oscillator PN is analyzed by calculating the effect of the cascade of the receiver and the transmitter oscillators, namely $e^{j\upsilon_R(n)}e^{-j\upsilon_R(n-\tau)}$, where τ is the relay processing delay. As the delay increases, the effect of the PN also increases, as is illustrated in Fig. 7.10a. From these curves, we can infer that for a delay $\tau < 10$ samples, the system can be considered free of ICI. More details can be found in [33]. On the other hand, Fig. 7.10b illustrates the loss due to nonlinear amplifier distortion and the PN for different processing delays. In this case, the effect of I/Q imbalance is not contemplated and an ideal ADC is assumed. The TX power is set at 0 dBm and $A_c = 50$ dB. Curves show that the aggregate effects of the PA and the PN reduce the performance of the system up to 16 dB.

The effect of ADC converter resolution is quantified in Fig. 7.11a. The antenna cancellation is $A_c = 40$ or 50 dB, whereas the digital cancellation is fixed to 50 dB. From the figure is observed that an ADC with low resolution severely affects the performance of the relay, and the effect is more severe for larger transmit power. The I/Q imbalance effects are illustrated in Fig. 7.11b. From this curve, it is possible to infer that in case of severe nonlinear distortion, the performance is determined by the PA. For a linear PA or large input back-off, the I/Q imbalance defines an error floor in the system performance.

7.3 ADC Resolution Requirements

After the downconversion, the automatic gain control (AGC) scales the received signal to fit into the dynamic range of the ADC. At this point, even after antenna and RF cancellation, the SI is tens of decibels above the signal of interest. Therefore,

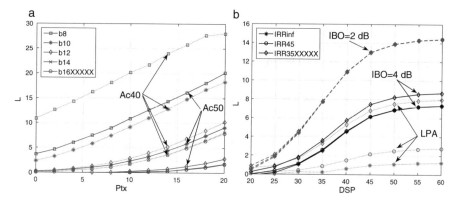

Fig. 7.11 (**a**) SNR loss curves: effects of transmitted power and ADC resolution. (**b**) SNR loss curves: effects of I/Q imbalances

high-resolution ADCs are required in order to avoid the signal of interest be buried in the quantization noise [34]. The relation between the cancellation level and the ADC resolution is shown in Fig. 7.11a, in Sect. 7.2.4. ADC resolution effects and transmitter PA nonlinearities are also addressed in [35] assuming the ADC as a memoryless nonlinearity using a classical approach [36].

Large bandwidth channels from hundreds of MHz to one GHz will be used in 5G networks to satisfy the demand of high data rates [37]. Then, high-speed ADCs are required to handle those broadband signals. On the other hand, it is well known that there is a compromise between the sampling rate (signal bandwidth) and dynamic range for practical ADCs [38]. Therefore, full-duplex (FD) operation in broadband channels is even more challenging, since high dynamic range and high sampling frequency are required.

High-resolution ADCs such as successive approximation register (SAR) ADCs are slow (the higher the resolution, the more clock cycles needed for the digital output word to be available) [39]. On the other hand, a TI-ADC structure offers two possible solutions: moderate resolution at high sampling rate or high resolution at a moderate sampling rate. In our case, where high effective resolution is required, a TI-ADC structure allows the use of these *slower-but-more-accurate* ADCs while keeping a moderate sampling rate adequate for the application. As digital dynamic range is proportional to the effective resolution, a TI-ADC allows sufficiently high dynamic ranges (an ENOB of 12 or above) that otherwise would be difficult to achieve, even through post-compensation on commercial ADCs [40]. As high-speed and high-resolution ADCs are a key component in broadband communications systems employing FD transceivers, TI-ADCs are a natural choice for analog-to-digital conversion.

A TI-ADC is an array of several ADCs working in parallel and interleaved in time by uniformly shifting their clocks, such that the overall conversion rate is proportionally increased. However, several mismatches between the ADCs (due to the manufacturing process) introduce nonlinear distortion that needs to be compensated for. The higher the ADC resolution, the higher the digital dynamic range available for post-processing, which is particularly important for FD communications. However, the usable dynamic range after AD conversion can be severely reduced due to distortion, measured both in terms of effective number of bits (ENOB) and spurious-free dynamic range (SFDR). A more complete description of TI-ADCs and mismatch compensation techniques can be found in Sect. 5.3.3.

In the following, the incidence of TI-ADC induced distortion in the performance of a full-duplex transceiver is analyzed. To reduce the harmful effects of the mismatches, we introduce a gain and offset mismatch compensation and analyze the effect of timing mismatch before and after digital correction [41]. For the simulation, a FD transceiver transmitting OFDM signals with 1024 subcarriers, a cyclic prefix of 64, and 16-QAM symbols is considered. The residual SI link (after antenna and RF cancellation) is a Rice channel of length 3 coefficients and a K-factor of 5 dB. The SR link is a time-varying frequency-selective Rayleigh channel of length 5 coefficients. The transmitted power is set to $P_{tx} = 20$ dBm unless it is stated otherwise. Antenna and RF cancellation are fixed to 50 dB and 30 dB, respectively. Additionally, we consider the challenging situation when the received signal is close to the receiver sensitivity level. We assumed that the signal of interest is received with a power of $P_{rx} = -80$ dBm, i.e., the SI is 30 dB over the signal of interest at the LNA input. We consider that the LNA gain is $K_{LNA} = \sqrt{P_{tx}P_{rx}/(A_{an}A_{rf}\mathscr{P}\{z(n)\})}$, where $\mathscr{P}\{\cdot\}$ is the peak-to-average power ratio, so that the received signal fits the ADC dynamic range, and it is considered that the AGC gain is also absorbed in K_{LNA}. We use the soft limiter model defined in Sect. 2.5.2 to describe the behavior of the PA, and consider an IBO of 8 dB to moderate clipping effects. The SNR at the FD receiver is given by $SNR = \frac{P_{rx}}{\sigma_R^2}$. The simulated ADC is an 8 branch TI-ADC with timing mismatch following the model in [42], working at a sample frequency equal to twice the signal bandwidth at baseband. We focus on timing mismatch because gain and offset are static errors and therefore straightforward to estimate and correct for, whereas timing mismatch is signal dependent and requires signal processing.

Different constellation size and coding rate will require different SINR thresholds to operate correctly. Systems with reduced constellation size are able to operate with low levels of SINR. If the size of the constellation increases, the SINR should be increased as well.

The figure of merit to analyze the incidence of the TI-ADC is SINR, evaluated after the DFT operation and channel equalization, and defined as:

$$SINR = \frac{1}{N} \sum_{k=0}^{N-1} \frac{E[|Z(k)|^2]}{E[|I(k)|^2]}, \tag{7.29}$$

where $E[|Z(k)|^2]$ is the power of the intended signal at subcarrier k and $E[|I(k)|^2]$ is the power of the interference signal introduced by non-ideal ADC and PA, the residual SI signal, and the channel noise.

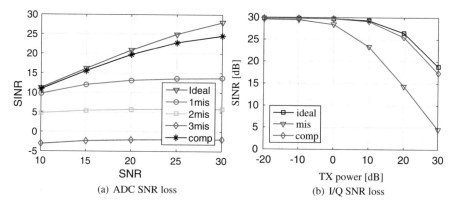

Fig. 7.12 (**a**) Receiver SINR for the ideal ADC, for a TI-ADC with 1–3% timing mismatch, and for the TI-ADC after compensation (Tx power = 20 dBm). (**b**) Receiver SINR for the ideal ADC, a TI-ADC with 1% timing mismatch, and for the TI-ADC after compensation when considering saturation effect on the PA (SNR = 30 dB, IBO = 8 dB)

From the results shown in Fig. 7.11a of Sect. 7.2.4, it is possible to infer that using less than 10 bits compromises seriously the performance of the SI cancellation. In Fig. 7.12a, the SINR degradation in the system with a time-mismatch of 0–3% in a 12-bit TI-ADC is investigated. From the figure, it is clear that after compensation the lost resolution due to the ADC distortion is almost completely restored. It is interesting to note that the effect of a timing mismatch greater than 2% is equivalent to having an ENOB of 8 bits, whereas a 1% mismatch is like having an ENOB of 10 bits. Therefore, compensation of ADC induced distortion is mandatory even for slight mismatches in the TI-ADC.

In Fig. 7.12b it is shown the SINR as a function of the transmitted power, for a fixed SNR and ADC resolution. It can be seen that, as expected, the higher the transmitted power, the lower the SINR, since the SI is larger for higher transmitted power.

The presented study shows that for a compensated TI-ADC, the SINR loss due to the FD operation mode is around 3 dB. In other words, this means that the SINR requirements at the FD receiver input are increased by 3 dB. For example, a half-duplex LTE user requires a SINR of 22.6 dB to operate with 64 QAM and coding data rate of 4/5 [43]. Using the proposed implementation, the 3 dB penalty increases the required SINR to 25.6 dB. On the other hand, a TI-ADC without compensation cannot reach the necessary SINR level, as noted from Fig. 7.12b.

7.4 Energy Efficiency and Spectral Efficiency

As discussed along the chapter, FD systems introduce a significant improvement in the spectral efficiency (SE). With the increasing interest in green information and communication technologies, energy efficiency (EE) has also gained great interest as a measure of performance of the system. Full-duplex relaying is usually used to increment SE and EE, since it increases the received SNR, and reduce the transmission power. Unfortunately, it is not always possible to maximize SE and EE at the same time. Moreover, in some cases the requirements to achieve the maximization of both figures are contradictory. The definitions of SE and EE in Chap. 3 need to be adapted to the FD relaying system, considering the network topology.

The relay operation mode (DF, AF, and CF) and the transmitting powers affect the SE and the EE. Therefore, the question about which mode is the best arises. For a relaying system with direct link, it is useful to define the SE gain as:

$$\Psi_{SE} = R_{\mathcal{M}} - R_{SD}, \tag{7.30}$$

where $R_{\mathcal{M}}$ is the rate of the relayed link using methods $\mathcal{M} = $ DF, AF, or CF, and R_{SD} is the rate of the direct link. In FD relayed links, increasing the relay power does not always lead to an improvement in the SE, since the SI also increases. Then, the EE can be defined as:

$$\Psi_{EE} = \frac{\Psi_{SE}}{P_t}, \tag{7.31}$$

where P_t is the total power, including power amplifiers, circuitry, SI cancellation. The EE allows to determine the amount of gain obtained per unit of consumed power.

A theoretical analysis based on independent Gaussian signaling concludes that DF is preferred if the relay has low transmitting power. Additionally, CF always provides cooperation gain regardless of the SI power. Using time sharing[1] between DF and CF, the SE can be improved. In the same manner, time sharing between FD and HD can be used to improve the SE, for some SI power conditions [18]. The method uses Gaussian signaling to arrive to more general conclusions, independent of the chosen modulation scheme.

Two-way relays are able to further increase the SE, since they are capable of simultaneously communicating with two end nodes, as described in Sect. 7.1.2. In this scenario, it is more convenient to define SE as:

$$\Psi_{SE} = R_{SD} - R_{DS}, \tag{7.32}$$

where R_{SD} and R_{DS} are the rates at the source and destination, respectively. For this scheme, the resulting average EE as a function of the SE presents two tendencies. Below a threshold, the EE increases with SE, since the SE dominates the EE. Above the threshold, the EE decreases with SE, since the total consumed power dominates the EE [44].

The escalation of energy consumption with frequency has been recognized as a major threat to the green implementation of mm-wave communications. This

---

[1] It is a technique where the system splits the communication time to use two transmission modes.

is because around 60 GHz, chips generally consume much more power than implementations working at lower frequency. As an additional challenge, the path loss is incredibly high at 60 GHz. For example, there is a drop of around 28 dB when compared to transmissions at 2.4 GHz. As a consequence, this kind of system is envisioned for short-range indoor communications. In this sense, the use of relays can be used to extend the network coverage and make the links more robust to blocks.

For short-range communications, the power amplifier (PA) consumption is comparable with that of the circuitry, due to the small chip size in indoor environment and the low drain efficiency of 60 GHz chips. The circuit power can be divided into static circuit power, P_{st}, and dynamic circuit power, P_{dy}, where the dynamic part takes into account the power consumption that depends on the throughput. The passive antenna cancellation does not consume additional power. On the other hand, the power consumed by RF and digital cancellation is non-negligible and can be considered static.

In this context, [45] analyze the EE of a FD DF relay at 60 GHz. Additionally, a cross-layer low-complexity resource allocation algorithm is proposed to maximize the EE, by setting the transmission power allocation, subcarrier allocation, and throughput assignment. The solution shows that EE-oriented algorithm makes trade-off between EE and SE, where the utilized transmission power may be lower than the available transmission power constraint, in order to maximize EE (see (7.31)). Therefore, the total power consumption could be lower than that for the SE maximization, and a higher EE can be obtained at the expense of SE.

7.5 Summary of the Key Points

Full-duplex (FD) systems enable the simultaneous co-channel transmission and reception in wireless environments. Besides doubling the theoretical spectral efficiency, these systems open a wide range of new applications, such as cognitive radio spectral sensing, more efficient medium access protocols, self-backhaul networks, among others. One of the main applications that benefit from the use of the FD technology are the relays.

The main impediment for the implementation of FD systems is the self-interference, an unavoidable coupling between the transmitter and receiver paths. To allow a practical implementation, the SI cancellation must be mitigated taking into account the hardware imperfections. The most important impairments are the PA nonlinearity, the quantization noise in ADC, and the I/Q imbalance. Consequently, several SI cancellation techniques for FD relays under these impairments were introduced and analyzed, showing the feasibility of FD communications in several contexts and their achievable performance.

References

1. M. Duarte, C. Dick, A. Sabharwal, Experiment-driven characterization of full-duplex wireless systems. IEEE Trans. Wireless Commun. **11**(12), 4296–4307 (2012)
2. T. Riihonen, R. Wichman, Energy detection in full-duplex cognitive radios under residual self-interference, in *9th International Conference on Cognitive Radio Oriented Wireless Networks and Communications (CROWNCOM)*, Oulu (IEEE, Piscataway, 2014), pp. 57–60
3. D. Korpi, T. Riihonen, V. Syrjälä, L. Anttila, M. Valkama, R. Wichman, Full-duplex transceiver system calculations: analysis of ADC and linearity challenges. IEEE Trans. Wireless Commun. **13**(7), 3821–3836 (2014)
4. M. Mohammadi, H.A. Suraweera, Y. Cao, I. Krikidis, C. Tellambura, Full-duplex radio for uplink/downlink wireless access with spatially random nodes. IEEE Trans. Commun. **63**(12), 5250–5266 (2015)
5. D.W.K. Ng, Y. Wu, R. Schober, Power efficient resource allocation for full-duplex radio distributed antenna networks. IEEE Trans. Wireless Commun. **15**(4), 2896–2911 (2016)
6. T. Riihonen, S. Werner, R. Wichman, Hybrid full-duplex/half-duplex relaying with transmit power adaptation. IEEE Trans. Wireless Commun. **10**(9), 3074–3085 (2011)
7. J.I. Choi, M. Jain, K. Srinivasan, P. Levis, S. Katti, Achieving single channel, full duplex wireless communication, in *Proceedings of the 16th Annual International Conference on Mobile Computing and Networking*, Illinois (ACM, New York, 2010), pp. 1–12
8. E. Everett, A. Sahai, A. Sabharwal, Passive self-interference suppression for full-duplex infrastructure nodes. IEEE Trans. Wireless Commun. **13**(2), 680–694 (2014)
9. K. Haneda, E. Kahra, S. Wyne, C. Icheln, P. Vainikainen, Measurement of loop-back interference channels for outdoor-to-indoor full-duplex radio relays, in *Proceedings of the 4th European Conference on Antennas and Propagation (EuCAP)*, Barcelona (IEEE, Piscataway, 2010), pp. 1–5
10. A. Raghavan, E. Gebara, E.M. Tentzeris, J. Laskar, Analysis and design of an interference canceller for collocated radios. IEEE Trans. Microw. Theory Tech. **53**(11), 3498–3508 (2005)
11. B. Debaillie, D.-J. van den Broek, C. Lavin, B. van Liempd, E.A.M. Klumperink, C. Palacios, J. Craninckx, B. Nauta, A. Pärssinen, Analog/RF solutions enabling compact full-duplex radios. IEEE J. Sel. Areas Commun. **32**(9), 1662–1673 (2014)
12. T. Riihonen, S. Werner, R. Wichman, Mitigation of loopback self-interference in full-duplex MIMO relays. IEEE Trans. Signal Process. **59**(12), 5983–5993 (2011)
13. D. Korpi, L. Anttila, V. Syrjälä, M. Valkama, Widely linear digital self-interference cancellation in direct-conversion full-duplex transceiver. IEEE J. Sel. Areas Commun. **32**(9), 1674–1687 (2014)
14. R. Lopez-Valcarce, E. Antonio-Rodriguez, C. Mosquera, F. Perez-Gonzalez, An adaptive feedback canceller for full-duplex relays based on spectrum shaping. IEEE J. Sel. Areas Commun. **30**(8), 1566–1577 (2012)
15. G. González, F. Gregorio, J. Cousseau, Blind self-interference cancellation for full-duplex relays, in *48th Asilomar Conference on Signals, Systems and Computers*, Pacific Grove (IEEE, Piscataway, 2014)
16. 5G CHAMPION deliverable D3.4: Algorithms for backhauling and fronthauling. 5G Communication with a Heterogeneous, Agile Mobile network in the Pyeongchang Winter Olympic competition (2017)
17. G.J. González, F.H. Gregorio, J. Cousseau, T. Riihonen, R. Wichman, Generalized self-interference model for full-duplex multicarrier transceivers. IEEE Trans. Commun. **67**, 4995–5007 (2019)
18. Z. Chen, T.Q.S. Quek, Y. Liang, Spectral efficiency and relay energy efficiency of full-duplex relay channel. IEEE Trans. Wireless Commun. **16**(5), 3162–3175 (2017)
19. T. Riihonen, Design and analysis of duplexing modes and forwarding protocols for OFDM(A) relay links. Ph.D. Thesis, Aalto University, School of Electrical Engineering, Department of Signal Processing and Acoustics (2014)

20. J.M. Laco, F.H. Gregorio, G. González, J.E. Cousseau, Patch antenna design for full-duplex transceivers, in *European Conference on Networks and Communications (EuCNC)*. IEEE Communications Society, European Commission (2017)
21. Analog Devices, GaAs HBT vector modulator HMC631LP3 1.8 - 2.7 GHz, v00.1007
22. T. Riihonen, R. Wichman, Analog and digital self-interference cancellation in full-duplex MIMO-OFDM transceivers with limited resolution in A/D conversion, in *46th Asilomar Conference on Signals, Systems and Computers*, Pacific Grove (IEEE, Piscataway, 2012)
23. F.J. Soriano-Irigaray, J.S. Fernandez-Prat, F.J. Lopez-Martinez, E. Martos-Naya, O. Cobos-Morales, J.T. Entrambasaguas, Adaptive self-interference cancellation for full duplex radio: analytical model and experimental validation. IEEE Access **6**, 65018–65026 (2018)
24. D. Korpi, L. Anttila, M. Valkama, Reference receiver based digital self-interference cancellation in MIMO full-duplex transceivers, in *IEEE GLOBECOM Workshops*, Austin (IEEE, Piscataway, 2014), pp. 1001–1007
25. E. Ahmed, A.M. Eltawil, A. Sabharwal, Self-interference cancellation with nonlinear distortion suppression for full-duplex systems, in *47th Asilomar Conference on Signals, Systems and Computers*, Pacific Grove (IEEE, Piscataway, 2013), pp. 1199–1203
26. L. Anttila, D. Korpi, V. Syrjälä, M. Valkama, Cancellation of power amplifier induced nonlinear self-interference in full-duplex transceivers, in *47th Asilomar Conference on Signals, Systems and Computers*, Pacific Grove (IEEE, Piscataway, 2013), pp. 1193–1198
27. A. Masmoudi, T. Le-Ngoc, A digital subspace-based self-interference cancellation in full-duplex MIMO transceivers, in *IEEE International Conference on Communications (ICC)*, London (IEEE, Piscataway, 2015), pp. 4954–4959
28. F. Gregorio, G. González, J. Cousseau, T. Riihonen, R. Wichman, Predistortion for power amplifier linearization in full-duplex transceivers without extra RF chain, in *IEEE International Conference on Acoustics, Speech and Signal Processing (ICASSP)*, New Orleans (2017)
29. A.C.M. Austin, A. Balatsoukas-Stimming, A. Burg, Digital predistortion of power amplifier non-linearities for full-duplex transceivers, in *IEEE 17th International Workshop on Signal Processing Advances in Wireless Communications (SPAWC)* (2016), pp. 1–5
30. R. Raich, H. Qian, G.T. Zhou, Orthogonal polynomials for power amplifier modeling and predistorter design. IEEE Trans. Veh. Technol. **53**(5), 1468–1479 (2004)
31. G. González, F. Gregorio, J. Cousseau, T. Riihonen, R. Wichman, Performance analysis of full-duplex AF relaying with transceiver hardware impairments, in *European Wireless (EW) Conference*, Oulu (IEEE, Piscataway, 2016)
32. G. González, F. Gregorio, J. Cousseau, T. Riihonen, R. Wichman, Full-duplex amplify-and-forward relays with optimized transmission power under imperfect transceiver electronics. EURASIP J. Wireless Commun. Netw. **2017**(1), 76 (2017)
33. T. Riihonen, P. Mathecken, R. Wichman, Effect of oscillator phase noise and processing delay in full-duplex OFDM repeaters, in *46th Asilomar Conference on Signals, Systems and Computers*, Pacific Grove (IEEE, Piscataway, 2012)
34. T. Riihonen, R. Wichman, Analog and digital self-interference cancellation in full-duplex MIMO-OFDM transceivers with limited resolution in A/D conversion, in *Proceedings of Asilomar Conference on Signals, Systems and Computers* (2012)
35. D. Korpi, T. Riihonen, V. Syrjälä, L. Anttila, M. Valkama, R. Wichman, Full-duplex transceiver system calculations: analysis of ADC and linearity challenges. IEEE Trans. Wireless Commun. **13**(7), 3821–3836 (2014)
36. J.J. Bussgang, Cross correlation function of amplitude-distorted Gaussian input signals. Res. Lab Electron., M.I.T., Cambridge, Tech. Rep. 216, vol. 3 (1952)
37. 3rd Generation Partnership Project; Technical Specification Group Radio Access Network, Study on scenarios and requirements for next generation access technologies; (release 14), 3GPP TR 38.913 V14.0.0 (2016)
38. S. Ouzounov, R. van Veldhoven, C. Bastiaansen, K. Vongehr, R. van Wegberg, G. Geelen, L. Breems, A. van Roermund, A 1.2v 121-mode ct #x00394; #x003a3; modulator for wireless receivers in 90nm CMOS, in *2007 IEEE International Solid-State Circuits Conference. Digest of Technical Papers* (2007), pp. 242–600

39. Y. Chi, D. Li, Z. Wang, A 16-bit 1ms/s 44mw successive approximation register analog-to-digital converter achieving signal-to-noise-and-distortion-ratio of 94.3db, in *2013 IEEE International Conference of Electron Devices and Solid-State Circuits (EDSSC)* (2013), pp. 1–2
40. C.A. Schmidt, O. Lifschitz, J.E. Cousseau, J.L. Figueroa, P. Julian, Methodology and measurement setup for analog-to-digital converter postcompensation. IEEE Trans. Instrum. Meas. **63**(3), 658–666 (2014)
41. C. Schmidt, G. González, F. Gregorio, J Cousseau, Compensation of ADC-induced distortion in broadband full-duplex transceivers, in *IEEE International Conference on Communications (ICC)*, París (IEEE, Piscataway, 2017)
42. M. Seo, M.J.W. Rodwell, U. Madhow, Comprehensive digital correction of mismatch errors for a 400-msamples/s 80-db SFDR time-interleaved analog-to-digital converter. IEEE Trans. Microw. Theory Tech. **53**(3), 1072–1082 (2005)
43. I.T.S. Sesia, M. Baker, *LTE - The UMTS Long Term Evolution: From Theory to Practiced* (Wiley, Hoboken, 2009)
44. H. Chen, G. Li, J. Cai, Spectral-energy efficiency tradeoff in full-duplex two-way relay networks. IEEE Syst. J. **12**(1), 583–592 (2018)
45. Z. Wei, X. Zhu, S. Sun, Y. Huang, Energy-efficiency-oriented cross-layer resource allocation for multiuser full-duplex decode-and-forward indoor relay systems at 60 GHz. IEEE J. Sel. Areas Commun. **34**(12), 3366–3379 (2016)

Chapter 8
Massive MIMO Systems

Abstract The development of massive multiple-input multiple-output (MaMIMO) techniques is motivated by the requirements of large spectral efficiency and reduced power consumption. The implementation of a large number of antennas offers a large spectral efficiency and link reliability. Moreover, the use of MaMIMO allows scaling down the transmitted power proportionally to the number of antennas used, which may lead to a significant improvement in terms of energy efficiency. Orthogonal frequency division multiplexing (OFDM) in combination with MaMIMO has a considerable potential to obtain very high data rates and high quality of service (QoS). However, MaMIMO systems require a mobile equipped with multiple antennas at the transmitter. This is a challenging issue in mobile devices mostly due to their size, cost, and computing power limitations. In MaMIMO, the radiated power per antenna decreases linearly with the number of antennas. Moreover, the effects of small-scale fading, non-coherent interference, and receiver noise are minimized. However, a massive number of antennas require a separate transceiver chain and power amplifier (PA) for each antenna (unless analog or hybrid analog-digital structures are used for beamforming purposes, in which case the number of RF chains can be reduced). In this situation, the size and costs of the analog front-end become a critical issue. The cost and size optimization implies the use of low-cost components which increase the imperfections that degrade the system performance. In this chapter, MaMIMO system performance, considering front-end RF imperfections and ADC/DAC with limited resolution, is carefully studied. Spectral and energy efficiency for the uplink and downlink scenarios are evaluated to quantify the overall system performance. Finally, low-resolution precoding techniques, antenna coupling, channel non-reciprocity, and channel estimation errors are also addressed.

8.1 Introduction

The requirements of communication systems with large spectral and energy efficiency motivate the development of MaMIMO. It offers large spectral efficiency and link reliability, and allows each transmitter chain to scale down the transmitted power proportionally to the number of antennas. This scaling potentially leads to a significant improvement in terms of energy efficiency [1, 2].

Massive MIMO consists of hundreds of antenna elements. Due to technological constraints, the implementation of multiple antennas in mobile devices is a challenging issue, and the use of large number of antennas is only affordable in base stations. In mobile devices only a few number of antennas are available [3]. The use of large number of antennas at the base station is essential to enhance the capacity without additional spectral resources [4, 5]. The combination of OFDM and MaMIMO has a huge potential to obtain very high data rates and high quality of service, and are promising candidates for future wireless cellular systems.

MaMIMO average out the effect of receiver noise, small-scale fading, and also the distortion generated by RF impairments [6]. The natural robustness against RF impairments is an excellent characteristic of MaMIMO that allows the implementation of RF chains with low cost and low power consumption components. This feature is fundamental in the search of power efficient and affordable wireless communication systems.

8.2   Single-Cell Massive MIMO System

In this chapter, a single-cell MaMIMO system based on OFDM is considered. It is composed of a base station with M antennas with a total power constraint P_{tr}, that serves L single-antenna users ($M \gg L$) operating in a time-division duplex (TDD) mode. It is assumed, for simplicity, that all users are equidistant in a radius r to the base station located at the cell center. Slow-fading Rayleigh channel model is assumed, and there is no coupling between antenna elements. The scenario is depicted in Fig. 8.1.

Fig. 8.1 MaMIMO scenario where a base station with M antennas communicates with L single antennas terminals

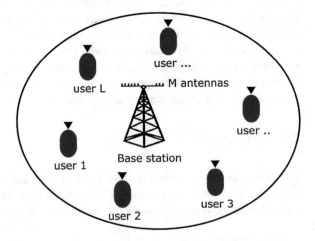

8.2.1 *Downlink*

The signals received by each user at subcarrier k, with $0 \leq k \leq N - 1$ where N is the OFDM symbol length, are grouped in the vector $\mathbf{y}_k$, defined by

$$\mathbf{y}(k) = \rho(k)\mathbf{H}(k)\mathbf{A}(k)\mathbf{s}(k) + \mathbf{w}(k), \tag{8.1}$$

where $\mathbf{y}(k) = [y^1(k), \ldots, y^L(k)]^T$, the wireless channel response at subcarrier k is modeled by the matrix $\mathbf{H}(k) = [\mathbf{h}^1(k), \ldots, \mathbf{h}^L(k)]^T$ of $L \times M$, with elements $h^{l,m}(k)$ for $1 \leq l \leq L$ and $1 \leq m \leq M$. $\mathbf{A}(k) = [\mathbf{a}^1(k), \ldots, \mathbf{a}^L(k)]$ is the precoding matrix for subcarrier k of size $M \times L$, with elements $a^{m,n}(k)$. The data symbols vector $\mathbf{s}(k) = [s^1(k), \ldots, s^L(k)]$, and $\mathbf{w}(k) = [w^1(k), \ldots, w^L(k)]^T$, where $w^l(k)$ are zero-mean white complex Gaussian processes with variance σ_w^2. The transmitted power is subject to a constraint given by $\rho(k) = \sqrt{p(k)}$, and it is assumed that the power per user is normalized, i.e., $E[\|s^l(k)|^2] = 1$. Additionally, the channel is assumed constant during an OFDM block frame (block fading) and its impulse response shorter than the cyclic prefix length.

8.2.1.1 Precoding Techniques

The downlink transmission requires a precoding operation that combines the messages to be transmitted to the L users and maps them into the M transmission antennas.

The use of nonlinear precoding techniques in systems that verify $M \gg L$ achieves the maximum system capacity. However, the practical implementation of nonlinear precoding techniques cannot be afforded for large M. Furthermore, the gain obtained by using nonlinear precoding methods comes at the cost of a considerable additional implementation complexity, which makes it impractical. This motivates the use of linear precoding techniques that provide a good performance with reduced implementation complexity.

To describe the different precoding techniques, initially it is assumed a perfect channel state information (CSI) at the MaMIMO base station (BS).

Considering a linear precoder, the transmit vector $\mathbf{x}_{(k)}$ is given as:

$$\mathbf{x}(k) = \sum_{l=1}^{L} \mathbf{a}^l(k)s^l(k) = \mathbf{A}(k)\mathbf{s}(k). \tag{8.2}$$

Then, the received signal at terminal l is given by

$$y^l(k) = \underbrace{\rho^l(k)\mathbf{h}^{lT}(k)\mathbf{a}^l(k)s^l(k)}_{\text{signal desired}} + \underbrace{\rho^l(k)\sum_{\substack{i=1 \\ i \neq l}}^{L} \mathbf{h}^{lT}(k)\mathbf{a}^i(k)s^i(k)}_{\text{Interference}} + \underbrace{w^n(k)}_{\text{noise}}. \tag{8.3}$$

A good selection of the weight vectors $\mathbf{a}^l(k)$ allows to minimize the multiuser interference, improving the signal-to-interference-plus-noise ratio (SINR). Considering a given channel realization, the SINR is given by

$$SINR^l(k) = \frac{p^l(k)|\mathbf{h}^{lT}(k)\mathbf{a}^l(k)|^2}{p^l(k)\sum_{\substack{i=1\\i\neq l}}^{L}|\mathbf{h}^{lT}(k)\mathbf{a}^i(k)|^2 + \sigma_w^2(k)}, \tag{8.4}$$

where $p(k) = \rho^2(k)$. The maximum achievable rate for user l at subcarrier k can be approximated by

$$R^l(k) \cong \mathrm{E}[\log_2(1 + SINR^l(k))]. \tag{8.5}$$

The total achievable rate of the system is the sum of the capacity of each user, $R_{\text{total}}(k) = \sum_{l=1}^{L} R^l(k)$ and the achievable average rate is $R_a(k) = \frac{R_{\text{total}}(k)}{L}$.

In the following, the existing linear precoding techniques are described.

Zero forcing (ZF) precoding is designed to achieve the zero multiuser interference (MUI). The precoding matrix for each subcarrier k is given by the pseudo-inverse of the channel gain matrix $\mathbf{H}(k)$,

$$\mathbf{A}_{ZF}(k) = \beta_{ZF}^{-1}\mathbf{H}^H(k)\left(\mathbf{H}(k)\mathbf{H}^H(k)\right)^{-1} \tag{8.6}$$

$$\beta_{ZF} = \sqrt{\|\mathbf{H}^H(k)\left(\mathbf{H}(k)\mathbf{H}^H(k)\right)^{-1}\|^2}. \tag{8.7}$$

Zero forcing precoding avoids the MUI and thus the system can be treated as a set of MIMO single user systems at each subcarrier k. However, the noise is amplified when the channel has large fades in its frequency response. In this case, particularly for frequency-selective channels, the noise severely degrades the system performance.

Maximum ratio transmission (MRT) precoding maximizes the signal-to-interference ratio (SIR) at the receiver. For each subcarrier k, MRT weights are derived by maximizing the ratio between the power of the desired signal and the power of the received signal subject to power restriction,

$$\mathbf{A}_{\mathrm{MRT}}(k) = \beta_{\mathrm{MRT}}^{-1}\mathbf{H}^H(k) \tag{8.8}$$

$$\beta_{\mathrm{MRT}} = \sqrt{\|\mathbf{H}^H(k)\|^2}. \tag{8.9}$$

Although the MRT precoder creates MUI, the residual MUI is minimized because channels of different users tend to be quasi-orthogonal for large number of antennas [1]. This scheme avoids matrix inversion leading to the simplest precoding solution.

8.2.2 Uplink

In the uplink scenario, it is considered that user terminals only weight their symbols by a power normalization factor. At the base station, the signal received at each antenna is linearly processed and decoded.

Considering an OFDM system, the signal received at the base station m-th antenna at the k-th subcarrier is the combination of the signal transmitted by all the user terminals.

$$y^m(k) = \rho_{UL} \sum_{l=1}^{L} h^{l,m}(k) x^l(k) + w^m(k), \tag{8.10}$$

where w_k^m is the receiver noise, $x^l(k)$ is the weighted symbol transmitted by user l, and $h^{l,m}(k)$ is the channel between user l and antenna m.

Assuming that the base station has knowledge of the channel response, it is possible to recover the symbols transmitted by each user. In the uplink scenario, the decoding process is implemented using the precoding matrices employed for data transmission in the downlink operation. The decoding matrix calculated using (8.8) is referred as maximum ratio combiner (MRC).

8.3 Precoding/Decoding Techniques with Imperfect Channel State Information

The precoder designs (8.6)–(8.8) are based on the assumption of perfect channel knowledge. However, in real systems the information at the transmitter is limited by multiple factors: (a) errors in the estimation due to pilot contamination, (b) delayed channel estimation, (c) channel reciprocity mismatch, and (d) limited feedback [7]. These factors influence the performance of precoders and must be contemplated.

The effect of imperfect CSI is considered using a simple model of hardware impairments [6]. Amplitude and phase deviations from the actual channel are modeled by

$$\tilde{\mathbf{H}}(k) = \mathbf{\Delta}(k) \odot \mathbf{H}(k), \tag{8.11}$$

where $\mathbf{H}(k)$ is the true channel, $\tilde{\mathbf{H}}(k)$ is the perturbed channel, $\odot$ is the Hadamard product, and $\mathbf{\Delta}(k)$ is the uncertainty matrix of $L \times M$, with elements defined as:

$$\Delta^{l,m}(k) = (1 + g^{l,m}(k)) e^{j\phi^{n,m}(k)}. \tag{8.12}$$

The amplitude error $g^{l,m}(k)$ and phase error $\phi^{l,m}(k)$ are independent zero-mean Gaussian random variables with variances σ_g^2 and σ_ϕ^2, respectively. The gain σ_g^2 and phase σ_ϕ^2 power errors are measured in dB and degrees, respectively. This type of multiplicative model of imperfect CSI includes most of the errors affecting the

MIMO channel model. Matrix $\mathbf{A}(k)$ is directly affected by $\tilde{\mathbf{H}}(k)$, according to the designed precoders (8.6)–(8.8) and normalization factors (8.7)–(8.9).

8.3.1 Channel Non-reciprocity and Antenna Coupling

There are unavoidable mismatches between transmitter and receiver chains that degrade the system performance. Downlink precoding in MaMIMO TDD systems assumes channel reciprocity, i.e., that uplink and downlink channels are identical. However, the end-to-end channels for the transmitter and the receiver are affected by non-symmetrical hardware impairments since radio chains of different characteristic are used for transmission and reception. E.g., high quality components are used in the BS, where implementation cost is not a restriction, but off-the-shelf components prone to errors are used in user terminals. On the other hand, mutual coupling effects between antenna elements cannot be prevented [8, 9]. The equivalent downlink and uplink channels between the BS and user l are a cascade of the transceiver front-end, the antenna mutual coupling at transmitter side, the physical channel, the antenna mutual coupling at the receiver, and the receiver front-end frequency response. In that model, it is assumed that L single antenna users are served by an M-antenna BS.

The equivalent channels of the downlink and the uplink are illustrated in Figs. 8.2 and 8.3, respectively.

Fig. 8.2 Equivalent downlink channel

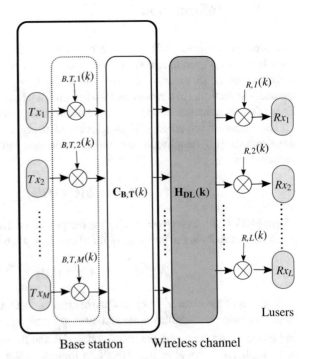

Fig. 8.3 Equivalent uplink channel

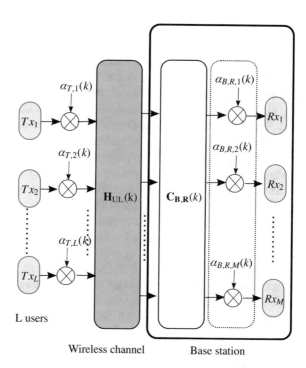

L users

Wireless channel   Base station

The channel considering a BS with M antennas and L single antenna users can be written as [10]:

$$\bar{\mathbf{H}}_{DL}(k) = \mathbf{A}_{U,R}(k)\mathbf{H}_{DL}(k)\mathbf{C}_{B,T}(k)\mathbf{A}_{B,T}(k) \tag{8.13}$$

$$\bar{\mathbf{H}}_{UL}(k) = \mathbf{A}_{B,R}(k)\mathbf{C}_{B,R}(k)\mathbf{H}_{UL}(k)\mathbf{A}_{U,T}(k), \tag{8.14}$$

where the antenna coupling at the base station is expressed by the $M \times M$ matrices, $\mathbf{C}_{B,T}$ and $\mathbf{C}_{B,R}$. $\mathbf{H}_{DL}$ and $\mathbf{H}_{UL}$ are the wireless channels, and assuming TDD verifies $\mathbf{H}_{DL} = \mathbf{H}_{UL}^{T}$.

The matrices $\mathbf{A}_{B,T}$ and $\mathbf{A}_{B,R}$ refer to the frequency response of the TX and RX chains, and are defined as:

$$\mathbf{A}_{B,T}(k) = \mathrm{diag}(\alpha_{B,T,1}(k), \ldots, \alpha_{B,T,M}(k)) \tag{8.15}$$

$$\mathbf{A}_{B,R}(k) = \mathrm{diag}(\alpha_{B,R,1}(k), \ldots, \alpha_{B,R,M}(k)), \tag{8.16}$$

where $\alpha_{B,T,m}(k)$, $\alpha_{B,R,m}(k)$ denote the response at subcarrier k of the transmitter and receiver chains in the m transceiver chain of the BS, respectively.

$\mathbf{A}_{U,T}$ and $\mathbf{A}_{U,R}$ denote the equivalent RX-TX mismatch matrices, and are given by

$$\mathbf{A}_{U,T}(k) = \mathrm{diag}(\alpha_{1,T}(k), \ldots, \alpha_{L,T}(k)) \tag{8.17}$$

$$\mathbf{A}_{U,R}(k) = \mathrm{diag}(\alpha_{1,R}(k), \ldots, \alpha_{L,R}(k)), \tag{8.18}$$

where $\alpha_{l,R}(k)$ and $\alpha_{l,T}(k)$ denote the response at subcarrier k of the receiver and transmitter chains of the l-th user.

In the following, the performance of ZF and MRT precodings without the assumption of channel reciprocity and antenna coupling is evaluated. It is assumed that the mutual antenna coupling exists only between neighboring antennas, i.e., $|\mathbf{C}_{B,R}|_{(i,j)} = 0$ if $|i-j| > 1$. The mutual coupling value, the frequency responses of the RX and TX chains, and the channel estimation error are modeled as zero-mean Gaussian process with respective variances ρ^2, δ_a^2, and σ_e^2. The effect of imperfect CSI is also included. Simulation parameters are summarized in Table 8.1.

Figure 8.4 illustrates the system capacity of MaMIMO system with $M = 200$ and 10 active users for ZF and MRT precodings. In these curves, the channel estimation error σ_e^2 is fixed to $-30\,\mathrm{dB}$, the TX-RX frequency response mismatch δ^2 varies from -30 to $-10\,\mathrm{dB}$, and the antenna coupling at the BS ρ^2 ranges from -30 to $-10\,\mathrm{dB}$.

It can be observed that channel estimation error affects the system performance. Moreover, high antenna coupling ($-10\,\mathrm{dB}$) and high levels of TX-RX mismatch also degrade the MaMIMO performance. In this case, in order to reduce the loss in terms of capacity, channel calibration techniques can be applied [11–13].

Capacity as a function of antenna coupling and the number of antennas in the base station is shown in Figs. 8.5 and 8.6. In these curves, it can be observed that increasing the number of antennas at the BS allows to reduce the SNR, i.e., decreasing TX power but maintaining identical data rate. This effect is even more pronounced for large antenna coupling. For large levels of mismatch, i.e.,

Table 8.1 Simulation parameters MaMIMO with antenna coupling

Parameter	Value
OFDM	$N = 512$
Users	$L = 10$
Base station	$M = 20\text{–}200$
Antenna coupling	$\rho = -30$ to $-10\,\mathrm{dB}$
TX-RX frequency response mismatch	$\delta^2 = -30$ to $-10\,\mathrm{dB}$
Channel estimation error	$\sigma_e^2 = -30$ to $-10\,\mathrm{dB}$

Fig. 8.4 Rate per subcarrier using ZF and MRT precoding. The rate was evaluated for a MaMIMO system with $M = 10$ to 200 and 10 active users

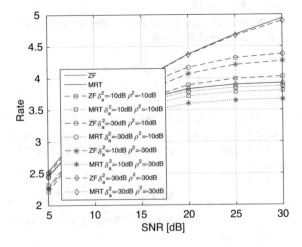

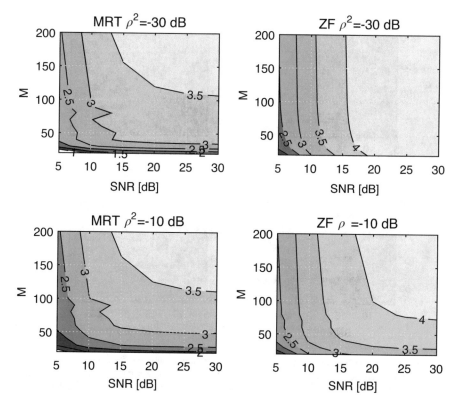

Fig. 8.5 Rate per subcarrier using ZF and MRT precoding varying the BS antenna coupling. The capacity was evaluated for a MaMIMO system with $M = 10$ to 200 and 10 active users. The channel estimation error is fixed to $\sigma_e^2 = -30$ dB and the TX-RX frequency response mismatch is $\delta^2 = -30$ dB

$\delta^2 = -10$ dB, illustrated in Fig. 8.6, the system capacity decreases and the harmful effect is even more pronounced when MRT precoders are employed.

8.3.2 Channel Estimation and Pilot Contamination

Channel state information is obtained at the BS using training sequences sent by each user. During the training, the L active users transmit orthogonal sequences and the BS estimates each channel. Then, assuming channel reciprocity, the uplink channel estimates are employed to create the precoding matrix.

Two channel characteristics are important to design the massive MIMO system and the channel estimation algorithm: (a) channel coherence bandwidth and (b) channel coherence time. The normalized channel coherence bandwidth is defined as the number of subcarriers for which the channel frequency response is constant, and is denoted as $N_{coh} = \frac{B_c}{\Delta_f}$, where $B_c = \frac{1}{T_d}$ is the channel coherence

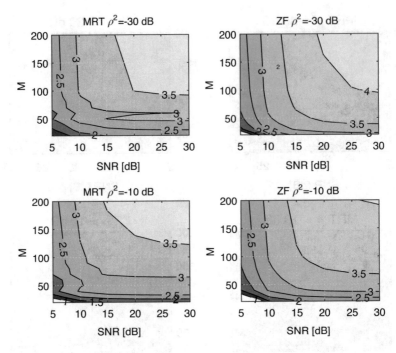

Fig. 8.6 Rate per subcarrier using ZF and MRT precoding varying the BS antenna coupling. The capacity was evaluated for MaMIMO system with $M = 10$ to 200 and 10 active users. The channel estimation error is fixed to $\sigma_e^2 = -30$ dB and the TX-RX frequency response mismatch is $\delta^2 = -10$ dB

bandwidth defined in Sect. 2.3.1, T_d is the channel delay spread, and Δ_f is the OFDM subcarrier spacing. The normalized channel coherence time defines the amount of OFDM symbols for which the channel is considered time-invariant, and is expressed by $\tau_c = \frac{T_c}{T_{symb}}$, where T_c is the channel coherence time, also defined in Sect. 2.3.1, and T_{symb} is the duration of an OFDM symbol. The base station is able to serve a limited number of users determined by the coherence time and coherence bandwidth of the channel, considering that a fraction τ_p of the coherence time is employed to transmit pilots, and the rest, denoted τ_d, is used to transmit information.

The BS can serve a maximum of [14]

$$L_{max} = \tau_p N_{coh} = \frac{T_p}{T_{symb}\Delta_f T_d} \tag{8.19}$$

users, where $T_p = \tau_p T_{symb}$ is the time employed to send pilot symbols. From this equation can be observed the dependence of the number of users to be served with the channel delay spread. To increase the amount of served users in case of channel with large delay spread, the time assigned to transmit pilots, T_p, can be increased. However, the energy spent on pilots is also increased reducing the energy efficiency of the system.

Fig. 8.7 Pilot contamination effect. During the training period, the BS of cell 1 estimates the channel using pilot transmitted by users k of cell 1 and cell 2

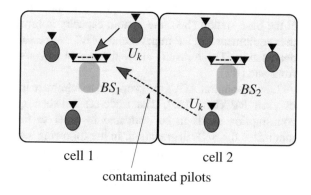

The most direct way to obtain channel estimations of different users is by employing orthogonal sets of pilot tones. However, in the case of multicell systems, where the same frequency band is reused in several cells, pilot contamination occurs [15]. If each cell is serving the maximum number of users, the BS will receive pilot tones contaminated by pilots transmitted by users of neighbor cells [14]. Consider a scenario of two cells as depicted in Fig. 8.7. Due to channel contamination, channel estimates calculated by the base station 1 are affected by the pilots transmitted in cell 2. In this case, the precoding matrix obtained from the contaminated channel estimates will generate an additional beam in the direction of the interfering user [16]. Pilot contamination limits the quality of channel estimate and therefore the ability to minimize the interuser interference. Pilot contamination decreases the expected gain of MaMIMO systems, and is considered one of the major impediment to its development [17, 18].

An alternative to mitigate the effect of channel contamination is the use of superposed pilots [19]. This technique uses an iterative data-aided channel estimation that outperforms the conventional time multiplexing of information and pilots, at the cost of an increase in the computational complexity. Other approaches have also been proposed to solve this problem in multicell scenarios [20–22]. The trade-off between additional complexity and obtained gain needs to be considered to choose the most adequate pilot decontamination technique for each case.

8.4 RF Front-End Minimum Requirements

A critical component of MaMIMO front-end, in terms of cost and power consumption, is the ADC due to the large number required when a large number of antenna elements are used.

The power consumption of the ADCs grows exponentially with the number of quantization bits and also grows with the sampling rates. Power consumption on the order of several watts are reported for ADCs with large resolution, e.g. 12 bits, and high sampling rate (wideband applications) [23], what is prohibitive for MaMIMO applications where hundreds of RF+ADCs chains are required.

From several works [24–26] it can be concluded that for large number of antennas at the base station (BS), the system capacity is limited by the impairments at the user equipment and RF impairments at BS can be supported with low performance degradation. These results allow the use of low-resolution ADCs at the receiver front-end (BS).

Low-resolution ADCs are proposed to minimize the cost and power consumption of each RF chain [27]. The trade-off between quantization noise and power consumption needs to be evaluated in order to find an optimal resolution that minimizes the SNR degradation. In the following section the use of low-resolution ADCs and its effect on the system performance, capacity, and effective SNR are addressed. Spectral and energy efficiency, both affected by the ADCs resolution, are studied in Sect. 8.5. The following sections are focused on the uplink context where the harmful effects of low-resolution ADCs are important.

8.4.1 Low-Resolution ADCs

Several aspects must be considered in the selection of the ADC. It is clear that there is a trade-off between the allowed quantization distortion and the consumed power. Moreover, the use of large signal bandwidth requires high sampling rate that inexorably increases the ADC power consumption. To evaluate the behavior of an ADC there are two fundamental parameters to be taken into account: (a) quantization noise and (b) power consumption.

Quantization Noise Based on the Bussgang theorem [28], the output of an ADC can be expressed by a scaling of the input signal and a distortion noise as $x_{adc} = Q(x) = \kappa_{ad}x + w_{ad}$, where $Q(x)$ denotes the quantization function, κ_{ad} is a scaling factor, and w_{ad} is the quantization noise.

Power Consumption The ADC power consumption can be modeled as:

$$P_{adc} = FoM2^{2b}f_s, \tag{8.20}$$

where FoM is a constant that depends on the ADC architecture and the technology, b is the resolution, and f_s is the sampling frequency.

The use of low-resolution ADCs affects in two ways the performance of MaMIMO systems:

- Receiver quantization noise: the quantization of the received information generates distortion, increasing the receiver noise floor that degrades the link throughput.
- Noisy channel estimation: The terminals transmit pilot symbols and, due to the ADC low-resolution, a noisy channel estimate is obtained.

The channel estimate can be expressed as:

$$\hat{\mathbf{H}}(k) = \mathbf{H}(k) + \mathbf{\Delta H}_q(k), \tag{8.21}$$

Fig. 8.8 MaMIMO uplink scenario where L users communicate with a base station with M antennas

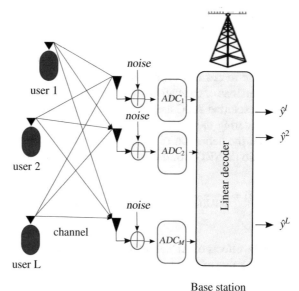

Base station

where $\Delta\mathbf{H}_q(k)$ is the channel estimation error due to quantization noise. A noisy channel estimate will affect the response of the downlink precoding matrix (transmission) and the transmitted signal will cause leakage over other users generating interuser interference.

On the other hand, the noisy channel estimates will also affect the decoding matrix and the spatial resolution of the MaMIMO system in the uplink scenario. The effect of noisy channel estimates affects the linear precoding matrices $\mathbf{A}(k)$ for MRC and ZF implementations. In this section the uplink scenario with L active users and a BS with M antennas is analyzed. This model is depicted in Fig. 8.8.

The split of channel estimates, Eq. (8.21), is reflected in the precoding matrices. The MRT matrix can be expressed as $\tilde{\mathbf{A}}_{MRC}(k) = \mathbf{A}_{MRC}(k) + \mathbf{A}_{MRC_q}(k) = (\mathbf{H}(k) + \Delta H_q(k))^H$, and the ZF precoder is given by $\tilde{\mathbf{A}}_{ZF}(k) = \mathbf{A}_{ZF}(k) + \mathbf{A}_{ZF_q}(k) = pinv(\mathbf{H}(k) + \Delta H_q(k))$.

The signal recovered at the BS for the user l at subcarrier k considering the modified precoding matrices can be expressed as:

$$\hat{y}^l(k) = \underbrace{\sqrt{\rho_u}\boldsymbol{a}^{lT}(k)\tilde{\boldsymbol{h}}(k)x^l(k)}_{\text{desired signal}\tilde{x}^l(k)} + \underbrace{\sqrt{\rho_u}\sum_{j=1,j\neq l}^{L}\boldsymbol{a}^{lT}(k)\tilde{\boldsymbol{h}}^j(k)x^j(k)}_{\text{MUI}(k)} + \underbrace{\boldsymbol{a}^{lT}(k)\tilde{\boldsymbol{n}}(k)}_{w_1(k)}$$

$$+ \underbrace{\boldsymbol{a}^{lT}(k)\boldsymbol{w}_{ad}(k)}_{w_{q_1}(k)} + \underbrace{\tilde{\boldsymbol{h}}(k)\sqrt{\rho_u}\boldsymbol{a}^{lT}(k)\tilde{\boldsymbol{h}}(k)\boldsymbol{x}(k)}_{\text{MUI}_q(k)} + \underbrace{\mathbf{a}_q^{lT}(k)\tilde{\boldsymbol{n}}(k)}_{w_2(k)} + \underbrace{\boldsymbol{a}_q^{lT}(k)\boldsymbol{w}_{ad}(k)}_{w_{q_2}(k)},$$

$$(8.22)$$

where the desired signal and several noise/interference contribution can be identified. The subscript q denotes a term that is generated by the low-resolution ADC. In case of an ideal ADC, these terms are zero. There are two sources of multiuser interference: (a) the MUI determined by the employed precoding method and (b) the MUI_q due to channel estimates affected by quantization noise. In Eq. (8.22), two scaled versions of channel Gaussian noise appear, denoted as w_1 and w_2. The effects of low-resolution ADC are given by the quantization noises, w_{q_1} and w_{q_2}.

Considering the desired signal and interference terms expressed in (8.22), and based on the definition of SINR (8.4), the signal to interference and quantization noise ratio (SINRQ) for the user l can be obtained as:

$$\text{SINRQ}^l(k) = \frac{|\tilde{x}^l(k)|^2}{|\text{MUI}(k)|^2 + |\text{MUI}_q(k)|^2 + |w_1(k)|^2 + |w_2(k)|^2 + ||^2 + |w_{q_1}(k)|^2 + |w_{q_2}(k)|^2},$$

(8.23)

where the effects of ADC low-resolution and channel estimation error are reflected. The uplink rate of user l can be calculated as:

$$R^l(k) \cong \text{E}[\log_2(1 + \text{SINRQ}^l(k))].$$

(8.24)

On the other hand, DAC resolution generates distortion in the precoding matrix and also affects the system performance [29, 30]. The issue of low-resolution DACs is not addressed in this chapter since the main focus is in low-resolution ADCs placed on RF chains of MaMIMO base stations and its effects over the system performance considering an uplink scenario. It is also considered that the terminals employ ADCs with enough resolution to avoid a significant increment of the receiver noise figure.

8.4.2 Performance Evaluation of a MaMIMO Uplink with Low-Resolution ADC

To evaluate the performance of MaMIMO systems with low-resolution ADCs, a simulation scenario that consists of a BS equipped with M antennas and L single antenna users, employing MRC and ZF linear precoders and decoders is presented. The uplink scenario is studied when the MaMIMO BS obtains the channel estimates from orthogonal pilot symbols transmitted by each user terminal.

The recovered symbol constellation at the BS for ADCs with 1 to 4 bits of resolution is illustrated in Figs. 8.9 and 8.10, for ZF and MRC decoding techniques, respectively. In the simulation, $L = 4$ active users and $M = 50$ antennas at the BS are considered. The recovered constellations with identical system parameters but $M = 250$ are plotted in Figs. 8.11 and 8.12. In these figures, ideal constellations are included as reference.

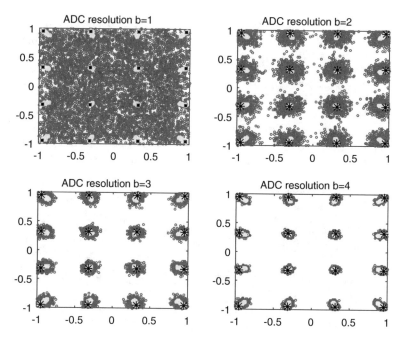

Fig. 8.9 Recovered symbol constellation using ZF decoding with 1–4 bits of ADC resolution with $M = 50$ and $L = 4$ active users (red markers for low-resolution ADC and yellow markers without quantization error)

Severe quantization noise can be observed when a 1 bit ADC is employed even for large number of BS antennas. Using the ZF decoder and 50 antennas, the residual MUI generated by quantization effects over channel estimates plus the receiver noise are severe and cannot be removed. The results are improved for the case of 250 antenna elements. When an ADC of 2 bits is adopted, the combination of quantization noise and interference is reduced.

In case of MRC, the performance is severally degraded due to quantization effects and multiuser interference even for large number of antennas. ADC resolutions larger than 4 bits are required for MRC receivers.

The rate of the system using ZF and MRC decoding as a function of a varying number of antennas at the base station and different ADC resolutions is illustrated in Fig. 8.13. It can observed that the rate depends on the ADC resolution and the number of antennas. The use of low-resolution ADCs requires an increment in the number of antennas to reach similar rate. The rate is also illustrated in Fig. 8.14, using ZF and MRC decoding with $M = 50, 100$, and 250 antennas.

A large number of antennas is useful to alleviate the quality requirements of the receiver chain components. It is verified that large levels of RF impairments are tolerated, which allows the implementation with low-cost off-the-shelf components. However, the use of a large amount of antennas requires more transceiver chains,

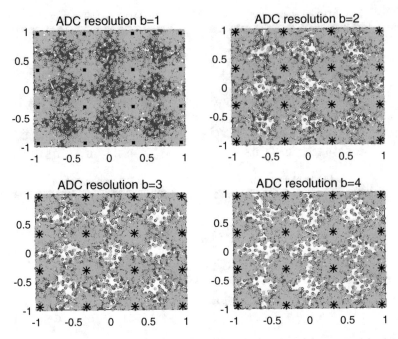

Fig. 8.10 Recovered symbol constellation using MRC decoding with 1–4 bits of ADC resolution with $M = 50$ and $L = 4$ active users (blue markers for low-resolution ADC and green markers without quantization error)

increasing the implementation cost. The trade-off between the allowed level of imperfections and the required number of antennas needs to be evaluated.

8.5   Power Consumption Analysis

The power consumption of communication systems was studied in Chap. 3. In this section, the MaMIMO scenario is presented and the contribution in terms of power consumption of each transceiver component is described.

Two typical blocks that can be identified in a communication system are the RF and the DSP blocks. The power consumption of RF chain has two dominant elements: the power amplifier and the ADC. The DSP block includes several elements that need to be considered when the power consumption of the overall system is under study. The decoding processing, which depends of the chosen codification technique, demands high computational complexity (number of operations) that results in large power consumption. Channel estimation and FFT/IFFT are also significant tasks to be executed by the digital processor [31].

Considering the uplink, the ADCs are one of the main sources of distortion and also the main power consumers. The ADC resolution defines its own power consumption and also the computational complexity of the baseband operations that are proportional to the word length. Channel estimation and linear decoding

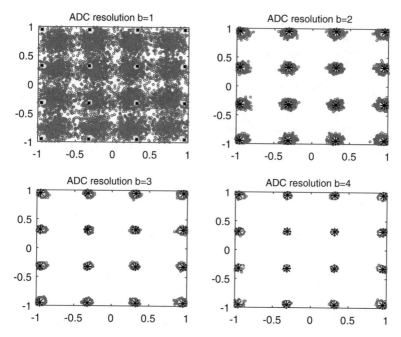

Fig. 8.11 Recovered symbol constellation using ZF decoding with 1–4 bits of ADC resolution with $M = 250$ and $L = 4$ active users (red markers for low-resolution ADC and yellow markers without quantization error)

are operations that depend on the ADC resolution b. The power consumption of the ADC and the DSP operations that depend on the ADC resolution are given by [24]

$$\tilde{P}_c(b) = 2M P_{adc}(b) + P_{lp}(b) + P_{ce}(b), \tag{8.25}$$

where P_{lp} and P_{ce} are the power consumed by linear processing and channel estimation, respectively. The number of operations required for channel estimation is given by

$$O_{ce}(b) = 2MLb\frac{BN_p}{T_cB_c} \tag{8.26}$$

where N_p is the length of the pilot sequence, B is the system bandwidth, and T_c and B_c are the time and bandwidth channel coherence, respectively. The power consumed for channel estimation is obtained dividing the number of operations by the computational efficiency L_{eff} of the employed DSP, i.e., $P_{ce} = \frac{O_{ce}}{L_{eff}}$.

The linear decoding process involves the following operations:

$$O_{lp}(b) = 2MLBb\left(1 - \frac{N_p}{T_cB_c}\right) \tag{8.27}$$

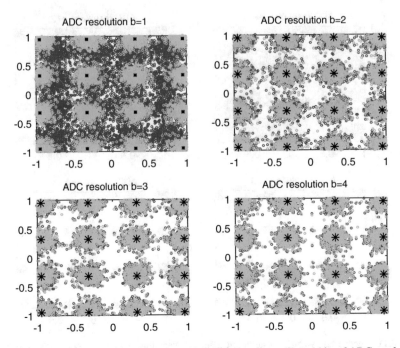

Fig. 8.12 Recovered symbol constellation using MRC decoding with 1–4 bits of ADC resolution with $M = 250$ and $L = 4$ active users (blue markers for low-resolution ADC and green markers without quantization error)

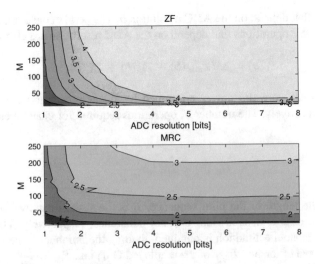

Fig. 8.13 User rate using ZF and MRC decoding. The rate was evaluated considering an ADC with a resolution varying from 1 to 8 bits and with a BS with $M = 10$ to 250 and $L = 4$ active users

Fig. 8.14 Rate using ZF
(blue line) and MRC (red
line) decoding. The capacity
was evaluated considering an
ADC with a resolution
varying from 1 to 8 bits and
with a BS with
$M = 50, 100, 250$ and 4
active users

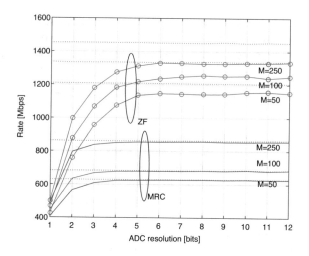

and the power consumed is $P_{lp} = \frac{O_{lp}}{L_{eff}}$. Replacing (8.20), (8.26), and (8.27) in (8.25), the consumption that depends on the ADC resolution is given by

$$\tilde{P}_c(b) = 2MFoM2^{2b}f_s + \frac{2MLBb}{L_{eff}}. \qquad (8.28)$$

The total power consumption of the uplink is obtained by adding the power required for signal transmission P_{tu} (single antenna users), the power consumption of RF chains of terminals P_U, and BS P_B [24]. They are given by

$$P_{tu} = \frac{L}{\rho_{pa}} p_{tx_u}$$

$$P_U = C_u L$$

$$P_B = C_B M, \qquad (8.29)$$

where p_{tx_u} is the power transmitted for each user, C_u and C_B are a set of constants that states for power consumption of RF chains of user terminals and base station, respectively.

The total power consumption of the system under study can be expressed by

$$P_c = \tilde{P}_c(b) + P_{tu} + P_U + P_B = 2MFoM2^{2b}f_s + \frac{2MLBb}{L_{eff}} + \frac{L}{\rho_{pa}} p_{tx_u} + C_u L + C_B M,$$
$$(8.30)$$

where the dependence of the overall power consumption with the ADC resolution, number of active users L, and number of BS antennas M is observed.

A metric to make a decision of the optimal solution between the ADC resolution and number of antennas at the BS is the energy efficiency (EE) measured in [bits/joules]. The EE can be defined for the uplink case as:

$$\eta_{EE}^{UL} = \frac{E[R_{UL}]}{P_c} = \frac{E[\sum_{l=1}^{L} R^l]}{P_c} = \frac{E[\sum_{l=1}^{L} log_2(1 + SINRQ^l)]}{\tilde{P}_c(b) + P_{tu} + P_U + P_B}, \qquad (8.31)$$

where R_{UL}, the total achievable rate of the system, is the sum of the rate of each terminal. From (8.31), it can be observed the dependence of EE with ADC resolution and the number of antennas. By increasing the ADC resolution, the effective SINRQ is improved, resulting in a larger user rate that increases the numerator of (8.31). However, a large resolution also increases the ADC power consumption and the consumption of the DSP block (linear processing and channel estimation) that results in a larger denominator of (8.31), reducing the EE. Similar results can be obtained by modifying the number of BS antennas M. A large M reduces the impact of ADC quantization noise and other impairments increasing the SINQR. On the other hand, the overall power consumption scales with M. There are a combination of ADC resolution and number of antennas that reaches the optimal results in terms of energy efficiency.

In the following, the EE of uplink MaMIMO system is studied using numerical evaluation. Table 8.2 resumes the simulation parameters employed in this section extracted from [31].

EE results are illustrated in Fig. 8.15 considering MRC and ZF decoders. It can be observed that EE degrades for low (<3) and high (>8) ADC resolution obtaining the optimal results for ADC resolution of 5–6 bits for ZF decoders and 4–5 bits for MRC implementations. There is also a large improvement when the number of antennas is scaled from $M = 20$ to $M = 40$. The duplication in the number of antennas creates an improvement in data rate that is larger than the growth of power consumption. However, when the number of antennas are increased to large values,

Table 8.2 Simulation parameters

Parameter	Value
Bandwidth B	10 MHz
ADC FoM	10^{-7}
PA efficiency ρ_{pa}	0.25
Power consumption terminals C_u	0.1 W
Power consumption BS C_B	0.1 W
Computational efficiency $L_{eff} B$	20,480
Users L	4
BS antennas M	10–250
Noise power	−90 dBm
Tx power	30 dBm
SNR	30 dB

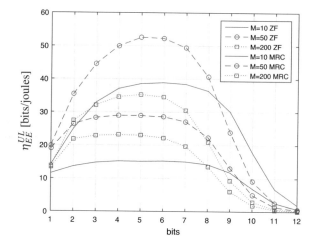

Fig. 8.15 Energy efficiency of MRC and ZF systems with an ADC resolution varying from 1 to 12 bits and with a BS with $M = 10, 50, 200$ and $L = 4$ active users

the increment in power consumption of the RF chains is not compensated with large rate gains, creating degradation of the system energy efficiency.

Surfaces plot are employed to illustrate the dependence of EE with ADC resolution and BS antennas. The results for ZF and MRC are shown in Figs. 8.16 and 8.17, respectively. In these figures, the maximum value of EE is indicated for both implementations. In case of ZF decoders, the maximum energy efficiency is reached using a BS with $M = 40$ antennas and ADC resolution of 6 bits. For MRC, 4 bits of resolution and 60 antennas are the optimal values. The low requirements in terms of number of BS antennas, 40–60, are justified by the fact that only $L = 4$ active users are considered. When the number of users is increased, it is demonstrated that a larger number of antennas are required [24].

It is also worth to mention that the use of large number of antennas is useful to mitigate the impairments of the RF front-end as is demonstrated for the case of ADC quantization noise. In this evaluation, only quantization noise is included. If other impairments are added, large number of antennas at the BS will be required to average out the harmful distortion.

The quantization noise employed in this section follows an additive model and its effects are similar to distortion noise generated by an LNA, inter-carrier interference (ICI) due to oscillator phase noise and carrier frequency offset all modeled by Gaussian noise (for large number of OFDM subcarriers). The mitigation of RF impairments provided by MaMIMO systems allows low-cost and low power consumption hardware that is mandatory when hundreds of receiver chains to be implemented.

Fig. 8.16 Energy efficiency
of ZF systems with an ADC
resolution varying from 1 to
12 bits and with a BS with
$M = 10$–250 and $L = 4$
active users

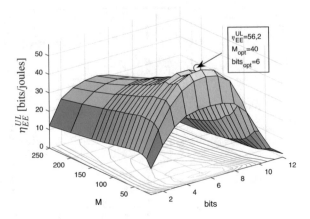

Fig. 8.17 Energy efficiency
of MRC systems with an
ADC resolution varying from
1 to 12 bits and with a BS
with $M = 10$–250 and $L = 4$
active users

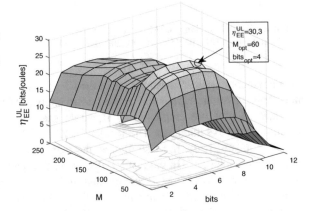

8.6   Summary of the Key Points

Massive MIMO is a technology with high potential to reach energy-efficient
wireless solutions. The use of large number of antennas minimizes the effect of the
RF front-end imperfections and allows the implementation of low cost/low power
consumption hardware. There are several issues that must be addressed to get a full
expansion of MaMIMO systems. Pilot contamination in multicell scenarios, channel
reciprocity, and antenna coupling are challenging topics to be investigated.

References

1. E.G. Larsson, O. Edfors, F. Tufvesson, T.L. Marzetta, Massive MIMO for next generation
   wireless systems. IEEE Commun. Mag. **52**(2), 186–195 (2014)
2. F. Rusek, D. Persson, B.K. Lau, E.G. Larsson, T.L. Marzetta, O. Edfors, F. Tufvesson, Scaling
   up MIMO: opportunities and challenges with very large arrays. IEEE Signal Process. Mag.
   30(1), 40–60 (2013)

3. L. Lu, G.Y. Li, A.L. Swindlehurst, A. Ashikhmin, R. Zhang, An overview of massive MIMO: benefits and challenges. IEEE J. Sel. Top. Signal Process. **8**(5), 742–758 (2014)
4. H.Q. Ngo, E.G. Larsson, T.L. Marzetta, Energy and spectral efficiency of very large multiuser MIMO systems. IEEE Trans. Commun. **61**(4), 1436–1449 (2013)
5. C.-S. Park, Y.-S. Byun, A.M. Bokiye, Y.-H. Lee, Complexity reduced zero-forcing beamforming in massive MIMO systems, in *Information Theory and Applications Workshop (ITA)* (2014), pp. 1–5
6. E. Björnson, J. Hoydis, M. Kountouris, M. Debbah, Massive MIMO systems with non-ideal hardware: energy efficiency, estimation, and capacity limits. IEEE Trans. Inf. Theory **60**(11), 7112–7139 (2014)
7. N. Jindal, MIMO broadcast channels with finite-rate feedback. IEEE Trans. Inf. Theory **52**(11), 5045–5060 (2006)
8. O. Raeesi, A. Gokceoglu, P.C. Sofotasios, M. Renfors, M. Valkama, Modeling and estimation of massive MIMO channel non-reciprocity: sparsity-aided approach, in *2017 25th European Signal Processing Conference (EUSIPCO)* (2017), pp. 2596–2600
9. O. Raeesi, A. Gokceoglu, Y. Zou, E. Björnson, M. Valkama, Performance analysis of multiuser massive MIMO downlink under channel non-reciprocity and imperfect CSI. IEEE Trans. Commun. **66**(6), 2456–2471 (2018)
10. Y. Zou, O. Raeesi, R. Wichman, A. Tolli, M. Valkama, Analysis of channel non-reciprocity due to transceiver and antenna coupling mismatches in TDD precoded multi-user MIMO-OFDM downlink, in *2014 IEEE 80th Vehicular Technology Conference (VTC2014-Fall)* (2014), pp. 1–7
11. J. Choi, Downlink multiuser beamforming with compensation of channel reciprocity from RF impairments. IEEE Trans. Commun. 63(6), 2158–2169 (2015)
12. J. Vieira, F. Rusek, F. Tufvesson, Reciprocity calibration methods for massive MIMO based on antenna coupling, in *2014 IEEE Global Communications Conference* (2014), pp. 3708–3712
13. H. Wei, D. Wang, H. Zhu, J. Wang, S. Sun, X. You, Mutual coupling calibration for multiuser massive MIMO systems. IEEE Trans. Wireless Commun. **15**(1), 606–619 (2016)
14. T.L. Marzetta, Noncooperative cellular wireless with unlimited numbers of base station antennas. IEEE Trans. Wireless Commun. **9**(11), 3590–3600 (2010)
15. O. Elijah, C.Y. Leow, T.A. Rahman, S. Nunoo, S.Z. Iliya, A comprehensive survey of pilot contamination in massive MIMO 5G system. IEEE Commun. Surv. Tuts. **18**(2), 905–923 (2016)
16. F. Rusek, D. Persson, B.K. Lau, E.G. Larsson, T.L. Marzetta, O. Edfors, F. Tufvesson, Scaling up MIMO: opportunities and challenges with very large arrays. IEEE Signal Process. Mag. **30**(1), 40–60 (2013)
17. K. Upadhya, S.A. Vorobyov, M. Vehkaper, Downlink performance of superimposed pilots in massive MIMO systems. IEEE Trans. Wireless Commun. **17**(10), 6630–6644 (2018)
18. E. Björnson, E.G. Larsson, M. Debbah, Massive MIMO for maximal spectral efficiency: how many users and pilots should be allocated? IEEE Trans. Wireless Commun. **15**(2), 1293–1308 (2016)
19. K. Upadhya, S.A. Vorobyov, M. Vehkapera, Superimposed pilots: an alternative pilot structure to mitigate pilot contamination in massive MIMO, in *2016 IEEE International Conference on Acoustics, Speech and Signal Processing (ICASSP)*, (2016), pp. 3366–3370
20. K. Upadhya, S.A. Vorobyov, An array processing approach to pilot decontamination for massive MIMO, in *2015 IEEE 6th International Workshop on Computational Advances in Multi-Sensor Adaptive Processing (CAMSAP)* (2015), pp. 453–456
21. H.Q. Ngo, E.G. Larsson, EVD-based channel estimation in multicell multiuser MIMO systems with very large antenna arrays, in *2012 IEEE International Conference on Acoustics, Speech and Signal Processing (ICASSP)* (2012), pp. 3249–3252
22. R.R. Müller, L. Cottatellucci, M. Vehkaperä, Blind pilot decontamination. IEEE J. Sel. Top. Signal Process. **8**(5), 773–786 (2014)
23. Texas instruments ADC products, http://www.ti.com/lsds/ti/data-converters/analog-to-digital-converter-products.page (2019)

24. M. Sarajli, L. Liu, O. Edfors, When are low resolution ADCs energy efficient in massive MIMO? IEEE Access **5**, 14837–14853 (2017)
25. D. Verenzuela, E. Björnson, M. Matthaiou, Hardware design and optimal ADC resolution for uplink massive MIMO systems, in *2016 IEEE Sensor Array and Multichannel Signal Processing Workshop (SAM)*, (2016), pp. 1–5
26. M. Sarajlic, L. Liu, O. Edfors, An energy efficiency perspective on massive MIMO quantization, in *2016 50th Asilomar Conference on Signals, Systems and Computers* (2016), pp. 473–478
27. Y. Li, C. Tao, G. Seco-Granados, A. Mezghani, A.L. Swindlehurst, L. Liu, Channel estimation and performance analysis of one-bit massive MIMO systems. IEEE Trans. Signal Process. **65**(15), 4075–4089 (2017)
28. J.J. Bussgang, Cross correlation function of amplitude-distorted Gaussian input signals, Res. Lab Electron., M.I.T., Cambridge, MA, Tech. Rep. 216, vol. 3 (1952)
29. S. Jacobsson, G. Durisi, M. Coldrey, C. Studer, Linear precoding with low-resolution DACs for massive MU-MIMO-OFDM downlink. IEEE Trans. Wireless Commun. **18**(3), 1595–1609 (2019)
30. C. Desset, L. Van der Perre, Validation of low-accuracy quantization in massive MIMO and constellation EVM analysis, in *2015 European Conference on Networks and Communications (EuCNC)* (2015), pp. 21–25
31. E. Björnson, L. Sanguinetti, J. Hoydis, M. Debbah, Optimal design of energy-efficient multi-user MIMO systems: is massive MIMO the answer? IEEE Trans. Wireless Commun. **14**(6), 3059–3075 (2015).

Chapter 9
Internet of Things

Abstract Internet of things (IoT) communications are considered as part of the last generation of wireless communication systems. As such, IoT represents a concept that leads to several design techniques to achieve different efficiency and performance objectives. In that scenario, base stations must be able to have an extended coverage for a massive number of low data rate nodes. On the other hand, IoT nodes are required to be low cost devices with restrictions on the total available power (i.e., battery operated) and processing capability. From the perspective of the nodes, the power constraints motivate the implementation of low cost, energy efficient devices operating with minimal transmission power and employing low constellation size. Regarding the base station extended coverage, depending on the frequency band used, several alternatives are considered. For the unlicensed ISM band, spread spectrum based or narrow-bandwidth based techniques are available, with very successful commercial products. For the licensed (cellular) bandwidths, there are also very interesting alternatives in terms of cost and coverage represented by: extended coverage-GSM (EC-GSM), machine-type LTE (LTE-M), and narrowband IoT (NB-IoT). Main characteristics of these techniques are related to the intensive use of repetitions and frequency diversity to obtain the extended coverage expected, the low cost premise to define the node devices and the compatibility (and simple software upgrade) with established cellular standards. Different from previous chapters, we use this chapter to illustrate design aspects of a particular IoT standard: NB-IoT. Emphasis is placed on its key features: frequency band used, cost, number of devices allowed, power consumption and capacity. To illustrate the performance of NB-IoT, we include the analysis of uplink RF impairment effects when LTE and NB-IoT coexist.

9.1 IoT Applications and Challenges

We made a brief introduction to general aspects of 5G specification and requirements in Chap. 1 . As stated, the main objectives are addressed to satisfy the different compromises of data rate, latency, mobility, and capacity of new and future services.

© Springer Nature Switzerland AG 2020
F. Gregorio et al., *Signal Processing Techniques for Power Efficient Wireless Communication Systems*, Signals and Communication Technology,
https://doi.org/10.1007/978-3-030-32437-7_9

We focus in this chapter on the particular communication characteristics demanded by devices directly operated from the network, and more generically: internet of things (IoT). When devices communicate through internet without requiring human interaction that is referred as IoT. The direct communication among these devices is known as Machine-Type-Communications (MTC).

Several different fields, such as communication technologies, microelectronics, data mining, big data handling, etc., reflect the advantages and advancements of IoT. A very brief summary of possible applications is the following,

- Smart grid—This corresponds to a number of applications related to the distribution and management of power, water, gas and heating. Smart metering is an appropriate example.
- Smart health—Where many applications related to automatic monitoring and tracking of patients, personnel and biomedical devices within hospitals are envisioned under the "smart health" category.
- Security and public safety—In this important category it is possible to include applications like home security, building access control, surveillance systems and other public safety mechanisms enabling *smart cities*.
- Asset tracking—Many applications that consider tracking and monitoring are visualized for IoT, ranging from monitoring the status of critical infrastructure (nuclear reactors, transport, bridges, etc.) and control industrial fleet management, wildlife monitoring and cattle tracking in agriculture, etc.
- Consumer devices and remote maintenance—Another vital application of IoT is in controlling sensors used for vending machine operations, vehicle diagnostics, remote home appliance control, etc.

Due to the remarkable and different amount of applications associated with machine-type communications, IoT is rather a family of communication techniques. In this family it would be necessary to take into account not only the resources of the physical link and the specific protocol, but also aspects related to the system architecture associated with the IoT application of interest.

In that sense, the IoT architecture required for the industrial use of IoT [1] (where different type of resources are monitored from a digital virtual replication) has nothing to do with that of a network of basic low power sensors (to measure, for example, a limited number of variables). Some key aspects that must be contemplated when going from human-to-human (H2H) to machine-type communications (MTC) are the following:

- Extended coverage: MTC devices can be located where the coverage of the network is very low (subsoil, underground parking, interior of buildings, etc.). Due to the restrictions of total available power and maximum power allowed for transmission in the channel (spectral mask restrictions), the MTC device cannot increase its power to reach the base station. This results in a very low signal-to-noise ratio operation in the base station, which is why it is necessary to develop mechanisms that improve performance in areas of low coverage.

- Massive number of devices: A key feature of IoT is the massive number of devices that may require access to the network. In the case of IoT cellular technologies it is foreseeable that the growth in the number of devices will be huge. In this way, managing a massive number of accesses to the network, while minimizing outage and providing adequate quality of service for different types of devices is a key requirement.
- Low power consumption: Since MTC devices do not require continuous data transmission and the amount of data per transmission is small, they do not need to be constantly connected or active. Also, IoT devices are generally low cost and low data rate with long-lasting batteries (10 years or more). Consequently, it is important that the different tasks associated with MTC be efficient in terms of energy consumption.
- High data rate—low latency vs low data rate—delay-tolerant devices: The current performance of communications networks is oriented to the demand of H2H devices, while in MTC the requirements are markedly different. By way of illustration, through the LTE network it is possible to offer high data rate and low latency videoconferencing services to a limited number of H2H users. However, the same network is not directly prepared to allow the efficient service of a massive number of low data rate delay-tolerant MTC devices associated with smart meter data.

Also, and although evident, the context of the application changes significantly, depending on the physical characteristic of the communications channel used. The essential characteristics are coverage (energy consumption) and defined bandwidth, although they could be others (complexity, etc.). We will focus here on basic aspects of the physical and medium access layers, describing some characteristics of simple and illustrative cases that allow to highlight the type of trade off that exists in this type of systems. In this sense we consider IoT as a generic way to describe low power wide area networks (LPWA).

After this introduction, we begin the chapter with a discussion (classification) of the different types of LPWA, putting emphases in unlicensed (proprietary) and licensed (cellular) IoT techniques, where we include some conceptual aspects related to obtain extended coverage. Next, we describe the specific physical and medium access characteristics of 3GPP NB-IoT. Later, we include some challenging aspects related to the physical layer of NB-IoT, mostly to cope with the problem of coverage and/or interferences. Finally, we conclude with a general discussion about main aspects to be taking into account in IoT design.

9.2   IoT Proprietary and/or Licensed Solutions

There are various standardization bodies, among them: IEEE, IETF, and 3GPP, that defined technologies to support IoT networks. LoRa, Sigfox, IEEE 802.15.4, IEEE 802.11, BLE, LTE-M, NB-IoT, etc. are some of the most well-known standards. An

Fig. 9.1 Bandwidth requirements vs. range for different IoT technologies. Source: Peter R. Egli, 2015, http://www.slideshare.net/ PeterREgli/lpwan

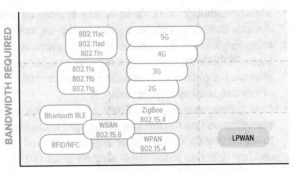

illustration of the different characteristics of available IoT technologies is depicted in Fig. 9.1. Taking into account not only the relative success of some applications in the market, but also the structural work carried out from the perspective of formal standards, is evident that the LPWA technologies evolve based on one of the two directions: by increasing the coverage of low power—short range devices, or by reducing the power of cellular technologies.

Following these ideas, a possible classification of the LPWA considers unlicensed (proprietary) networks and licensed networks. Examples of LPWA in non-licensed bands are LoRa, Sigfox, etc. These networks are designed for the industrial, scientific, and medical (ISM) band and are available for various applications, although the associated standards are only backed by industrial alliances. The interesting aspect related to these IoT systems is the availability (one of the first IoT networks with good coverage in many countries is Sigfox. Also LoRa is popularly accepted since the last years). Since they operate at the ISM band, one drawback of the aforementioned technologies is that they are susceptible to interference, and they unable to meet QoS requirements. On the other hand, licensed LPWA are based on standards based on 3GPP, such as EC-GSM, LTE-M, and NB-IoT [2]. From this perspective, the IoT design effort is mostly addressed to adequate previous standards to new MTC applications, i.e., reducing transceiver design complexity and decreasing power consumption (or increasing coverage).

In the following paragraphs, we provide a brief description of different IoT technologies, either developed for unlicensed bands or for 3GPP licensed bands. Basic concepts related to key IoT system coverage design are discussed next. Further details regarding NB-IoT will be introduced later, in the following section.

- Unlicensed—low energy—low range

  - **IEEE 802.15.4** Wi-Smart Utility Networks consortium, established in 2011, uses the IEEE 802.15g/e for the physical (PHY) and medium access (MAC) layers to enable efficient management of utility services, i.e., water, gas, electricity, etc. It used sub-GHz and 2.4 GHz ISM bands. The standard PHY layer supports three different formats: multirate frequency shift keying (MR-FSK), MR-offset quadrature phase shift keying (MR-OQPSK), and MR-OFDM for higher data rates (i.e., from 4.8 up to 800 kbps). In this way is able to accommodate thousands of users. This is the umbrella standard for many other IoT technologies which have defined higher layers. That includes direct connection to Internet called 6LoWPAN, which simplify implementation to send and receive data from the cloud. Zigbee is another IEEE 802.15.4 popular standard for low power, low data rate wireless networks. It supports tree, star, and mesh networking to low data rate networks spread over a larger area. The price for this may be shorter battery life for devices that serve as repeaters.

– **IEEE 802.11ah** To support IoT applications, IEEE 802.11 family introduced the IEEE 802.11ah amendment (Ha-Low) in the Sub-1 GHz band. The main objective of this technology is to fill the gap of wide personal area networks (WPAN) and LPWA networks by providing adequate coverage (up to 1 km) and high data rates (peak data rates exceeding 100 kbps). IEEE 802.11ah facilitates the support of various wireless technologies, by acting as backhaul network. The physical layer (PHY) adopts OFDM that transmit at the rate 10 times slower than IEEE 802.11ac, an earlier standard, so that extending the communication range. IoT devices are enabled with mechanisms to save energy during the inactive periods but yet retain their connection/synchronization with the access points.

– **Bluetooth low energy (BLE)** Bluetooth PHY uses GFSK modulation, and frequency hopped spread spectrum (FHSS), but in BLE the modulation is based on direct sequence spread spectrum (DSSS) modulation which have some interference avoidance features. BLE is very common in wireless peripherals for laptops and cell phones, such as wireless mouses or wireless headsets, but it is also the most common wireless standard for fitness bands and many wearable IoT devices. Bluetooth uses the 2.4 GHz ISM band. It was originally defined in IEEE 802.15.1, but now it is managed by the Bluetooth Special Interest Group (SIG) which is an alliance of companies making Bluetooth devices. Over time, Bluetooth IoT standards have diversified, adding Bluetooth Low Energy (BLE or Bluetooth Smart) in 2006 and Bluetooth 5 in 2016.

• Unlicensed—low energy—long range (see Table 9.1)

– **Sigfox [3]** is a LPWA technology that operates in an ultra-narrow band, i.e., 100 Hz sub-GHz ISM band carrier and uses binary phase shift keying (BPSK) modulation. The use of the ultra-narrow band results in decreased noise levels, improved receiver sensitivity, and low cost antenna design. The coverage is 30–50 km in rural areas and 3–10 km in urban areas (maximum path loss of 162 dB at the European 868 MHz and US 902 MHz ISM frequencies). However, the data rate provided is limited to 100 bps. This technology uses unslotted ALOHA as the MAC layer and it incorporates encryption and/or forward error correction (FEC) for the data packets only in recent versions.

Table 9.1 Characteristics of typical unlicensed bandwidth, low energy—long range technologies

	LoRa	SigFox	Ingenu
Modulation	Chirp spread spectrum	Ultra-narrow band DBPSK (UL), GFSK (DL)	Random phase multiple DSSS (UL), CDMA (DL) access
Bandwidth	125 kHz	160 Hz	1 MHz
Frequency bands	Sub-GHz ISM	Sub-GHz ISM	ISM 2.4 GHz
Data rate	0.3–27 kps (50 kps with FSK)	100 bps (UL), 600 bps (DL)	156 kbps (UL), 624 kbps (DL)
Coverage	5 km (urban), 15 km (rural)	10 km (urban), 50 km (rural)	5–6 km (urban)
Ner. of channels	10 in EU 64 + 8 (UL) and 8 (DL) in US	360 channels	40 × 1 MHz channels up to 1200 signals/channel
Link symmetry	Yes	No	No
FEC	Yes	No	Yes
MAC	Unslotted ALOHA	Unslotted ALOHA	CDMA similar
Adaptive data rate	Yes	No	Yes
Payload length	up to 250 B	12 B (UL), 8 B (DL)	10 kB
Multicast updates	Yes	No	Yes
Localization	Yes	No	No

- **LoRa** is also a proprietary LPWA technology [4]. It operates on the sub-GHz ISM band and uses an unslotted ALOHA based MAC. It offers also data encryption and FEC. For the physical layer, it uses a chirp spread spectrum (CSS) modulation, which results in noise and interference resilient signals. LoRa offers data rates up to 50 kbps. The upper layers and the system architecture of LoRa are being standardized by the LoRa Alliance under Long Range Wireless Access Network (LoRaWAN) specification [5], which supports different classes of IoT devices based on latency. The LoRa Alliance claims to provide a maximum coupling loss of 155 dB in the European 868 MHz band, and 154 dB in the US 902 MHz band [5].

- **Ingenu** Is another example of an LPWAN with their Random Phase Multiple Access (RPMA) technology, this time using the 2.4 GHz license exempt band. RPMA is a DSSS-based modulation complemented by a pseudo random time of arrival that helps distinguishing users multiplexed on the same radio resource. Ingenu claims to achieve an maximum path loss (MPL) of 172 dB in the United States and 168 dB in Europe [6]. Its underlying technology uses more power than others, so it is not as well suited for battery-powered applications.

- Licensed—low energy—long range (see Table 9.2)

  - **Enhanced coverage GSM (EC-GSM)** aims to provide long distance and low power consumption communications among IoT devices to coexist with the existing wireless networks. This standard is built on the basis of the Enhanced

Table 9.2 Characteristics of 3GPP low energy—long range technologies

	eMTC (LTE Cat M1)	NB-IoT	EC-GSM
Deployment	In-band LTE	In-band, guard-band, standalone	In-band GSM
Coverage (MCL)	155 dB	164 dB	164 dB, 33 dBm power class. 154 dB, 23 dBm power class
Downlink	OFDMA, 15 kHz spacing Turbo coding, 16 QAM	OFDMA, 15 kHz spacing	TDMA/FDMA, GMSK and 8PSK
Uplink	SC-FDMA, 15 kHz spacing, Turbo coding, 16 QAM	Single tone: 15 kHz and 3.75 kHz spacing. SC-FDMA: 15 kHz spacing, turbo code	TDMA/FDMA, GMSK and 8PSK
Bandwidth	1.08 MHz	180 kHz	200 kHz
Peak rate (DL/UL)	1 Mbps for DL and UL	DL: 127 kbps [7], UL: 158 kbps [7]	DL and UL (4 timeslots): 70 kbps (GMSK), 240 kbps (8PSK)
Duplexing	FD and HD, FDD and TDD	HD, FDD	HD, FDD
Power saving	PSM, ext, I-DRX, C-DRX	PSM, ext. I-DRX, C-DRX	PSM, extended I-DRX
Power class	23 dBm, 20 dBm	23 dBm	33 dBm, 23 dBm

General Packet Radio Service (eGPRS), which is an improved technique allowing for higher data rates than traditional GSM networks. The constraint in GSM to access a massive number of devices was markedly improved, and battery life of IoT devices based upon this standard can be prolonged to up to 10 years. It introduces a 10 dB range improvement with respect to GSM coverage.

– **3GPP LTE-M** Two types of MTC scenarios were initially defined for the evolution of 3GPP: device to server links and device to device links based on the cellular network. Right from early stages it was realized that the main challenges for MTC services were the signaling overhead and network congestion due to the massive number of devices. Therefore, 3GPP focused on the overhead in the control signals. One of the first schemes considered and described in 3GPP for addressing the substantial number of devices contending for a preamble, is the Access Class Barring (ACB). A device number generated randomly is compared against an access probability number that is transmitted by the device. If the device generated number is higher than the access probability, then a device will access the channel. To further reduce complexity and power consumption, a new device category (namely UE category 0), is introduced to allow the support of single antenna. The improved efficiency in power consumption, comes at the cost of worse performance (approx. 5 dB). Only the minimum LTE carrier bandwidth, i.e., 1.4 MHz with 6 physical resource blocks (PRB) of 180 kHz is supported. Moreover, repetition in downlink (DL) and uplink (UL) shared channel subframes is used to improve coverage. Also, power saving mode (PSM) allows devices to remain registered in the network even when they cannot be reached by the network.

- **Narrowband IoT (NB-IoT)** is an LPWA technology introduced in 3GPP Rel-13, compatible with previous LTE. Operating on a 180 kHz channel bandwidth, NB-IoT is characterized by peak data rates of less than 200 kbps, limited mobility, an even lower power consumption and component cost. NB-IoT only supports FDD (unlike LTE-M) and supports three types of communications: in-coverage, outside-coverage by using GSM carriers, and guard-band, where NB-IoT channel is placed in the guard band of LTE channels. The narrow bandwidth enables multiplexing more users in the same bandwidth in UL direction. On the other hand, the minimum resource allocation unit is set to one subcarrier for NB-IoT nodes. Moreover, two subcarrier spacing are supported in UL: short (3.75 kHz) and the one used also in DL (15 kHz). Frequency hopping is also supported in the device random access channel. To meet the requirements for extended battery life (over 10 years), discontinuous reception (DRX) cycle for NB-IoT devices, is extended from 2.56 s to approximately 3 h. It introduces a 10 dB range improvement with respect to LTE-M and 20 dB related to 3G GPRS coverage.

Just to be fair, there is not such exclusive division between licensed and unlicensed technologies for IoT. Some ideas developed for LTE Rel. 13 [8], to unload the licensed bands with those of ISM bands, can also be used with IoT. With MTC characteristics in mind, an *unlicensed* NB-IoT is now available [9, 10], where the same benefits of licensed NB-IoT can be obtained (with particular modifications to satisfy the ISM requirements) but around 900 MHz ISM band.

In any case, we are mostly interested in the low energy long range kind of applications for IoT since they introduce very interesting challenges in terms of system and device designs. So, that will be the focus in the following discussions where some basic concepts related to IoT system coverage design are discussed.

9.2.1 Coverage and Capacity: Shannon and Transmission Bandwidth

To obtain an extended coverage, the receiver node must have a very low sensitivity. Sensitivity is defined as the minimum power needed to obtain an arbitrary error rate, and can be expressed (dB) as

$$P_r = SNR_r + 10\log(N_0 W) \tag{9.1}$$

where SNR_r is the signal-to-noise ratio (dB) required for the specified error rate (that can also include the noise figure (NF) of the receiver), W is the signal bandwidth (Hz) and N_0 is the noise power spectral density. We have at hand two possible approaches to obtain low sensitivity: reduction of the signal bandwidth W, or working at a low SNR regime using some specific transmission—reception technique.

The first approach leads to the use of narrowband signals. This is the strategy chosen by Sigfox. Even if the SNR required for the error rate specified remains high, the narrowband signal allows to obtain low sensitivity.

The second approach considers a noise reduction for the error rate considered. A commonly used technique to this purpose is the repetition factor, i.e., repeating the signal ℓ times leads to a processing gain equal to $10 \log(\ell)$ dB at the receiver. Among others efficient alternatives, this is the basic approach considered by 3GPP IoT techniques. Other techniques may also be envisaged, such as the use of channel coding, which reduces the bit rate but also the effective SNR per bit E_b/N_0, required for the same error rate. Spread spectrum techniques can be considered in this application to reduce the effective SNR using the processing gain (related to the bit rate—symbol rate ratio). LoRa and other techniques basically are following this approach.

Shannon defined the maximum achievable bit rate for an AWGN channel, i.e., the capacity C as

$$C = W \log_2 \left(1 + \frac{P}{N} \right) \tag{9.2}$$

where $P = E_b B$ is the signal power, $N = N_0 W$ is the noise power, and B is the transmission bandwidth. Note that, at an extreme coverage scenario, $P/N \ll 1$ and previous equation becomes

$$C \approx \frac{P}{N} \log_2 e \tag{9.3}$$

that allows to conclude that system capacity in that regime only depends on P and N but not the signal bandwidth. However, for the sake of the spectral efficiency, it is better to use a small signal bandwidth in low SNR scenarios.

Based in (9.2) it is possible to obtain the maximum spectral efficiency achievable for a given E_b/N_0. The maximum spectral efficiency $\eta_{max} = C/W$, is given by

$$\eta_{max} = \log_2 \left(1 + \frac{E_b}{N_0} \eta_{max} \right) \tag{9.4}$$

To illustrate the different trade-offs, Fig. 9.2 depicts [11] a comparison of the spectral efficiency vs E_b/N_0 obtained with four IoT solutions: Sigfox, based on narrow-band signals (following the specification translates to an specific point in the graphic); LoRa, based on orthogonal modulations (where increasing spreading

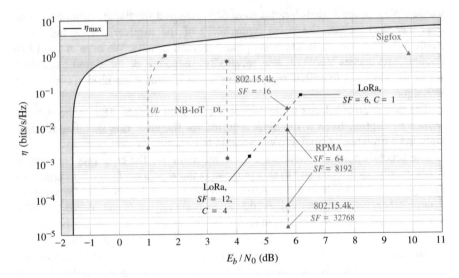

Fig. 9.2 Spectral efficiency of different IoT systems [11]

factors allow to improve coverage but also compromises bandwidth); RPMA, based on the standard 802.15.4 K that uses an specific spread spectrum technique and repetition factor; and the Narrow-Band IoT, that uses narrowband and an efficient turbo coding. For all technologies the Binary Error Rate (BER) is equal to 10^{-5}.

It can be seen that Sigfox technology, since it adopts the narrow-band approach, can be distinguished from other solutions. Even when, for a fair comparison, an equivalent higher BER could be setup for NB-IoT (since it works in a licensed band of controlled interference against the other techniques that work in the ISM band), this technology is very close to the limit of Shannon.

9.2.2   Maximum Coupling Loss and Maximum Path Loss

In order to obtain certain coverage, in addition to the payload or information channel reaching a certain minimum throughput, also the control channels need to be operable (to synchronize the device to the network, to acquire the associated control channels, etc.). Also, it is important that the coverage is balanced. When comparing the coverage of different logical (information and control) channels, the balance is easily done in the same direction (UL or DL) using similar assumptions of transmit power and thermal noise levels at the receiver. When we are looking to balance the UL and DL, assumptions need to be made on device and base station transmit levels, and also about level of thermal noise that can typically be expected to limit the performance.

Multiple factors impact the thermal noise level at the receiver: temperature, bandwidth (the larger the bandwidth of the signal, the higher the absolute noise power), and the noise figure (NF). Lowering the NF would improve the coverage but also implies a more complex and costly implementation. An alternative to improve coverage is to increase the transmit power in the system, but this would imply an increase in implementation complexity and cost. Also the bandwidth needs to be kept low to minimize implementation and impact to existing deployments.

Maximum coupling loss (MCL) is a common measure to describe the coverage that a wireless link can satisfy. It is the limiting value of the coupling loss at which a service can be delivered, and therefore defines the coverage of the service. It is used instead of a measure of distance, that depends on the carrier frequency and the environment (i.e., indoor, outdoor, urban, sub-urban, and rural). In theory, it can be defined as the maximum loss in the conducted power level that a system can tolerate and still be operational (defined by a minimum acceptable received power level). MCL can be calculated as the difference between the conducted power levels measured at the transmitting and receiving antenna ports as the reference point. The directional gain of the antenna is not considered when calculating MCL.

MCL is defined by

$$MCL(dB) = P_{T_x} - (NF + SINR + N0), \tag{9.5}$$

where

- Maximum transmitter power P_{T_x}: For DL MCL calculation, this is the power amplifier (PA) power of the base station. For UL MCL calculation, the PA power of the IoT node is used. Usually IoT nodes support different power classes (for example, 3GPP LTE-M supports two UE power classes: a 23 dBm power class, and a new 20 dBm power class).
- Receive Noise Figure NF: Similarly to P_{T_x} that is based on the PA, the noise figure is based on the low noise amplifier (LNA) of the receiver front-end. The front-end insertion loss, quality, and current draw of the LNA can affect the NF and so typically the NF for the IoT nodes are higher than for the base station (which generally has less concerns with respect to power consumption and cost).
- Bandwidth of Signal B: This is the bandwidth of the actual signal transmitted, not the bandwidth of the system (for example, for LTE-M, if 2 physical resource blocks (PRBs) are used, then 2×180 kHz is used, not the full system bandwidth).
- $N0$ is the specified noise floor (noise power spectral density N_o at the specified bandwidth B, i.e., $N0 = N_0(dB/Hz) + 10 \log B$)
- Required SINR $SINR$: This value is a measure of how much noise the design (e.g., modulation, coding rate, coding type, transmission mode, and/or diversity scheme) can tolerate and still work within a certain performance. The performance metric is often block error rate (BLER) but can be also acquisition time or speed.

Fig. 9.3 Maximum coupling loss and maximum path loss illustration

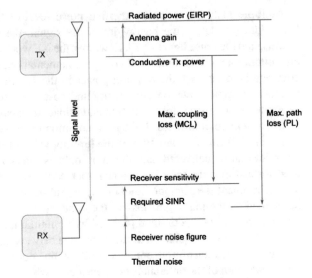

As stated before in Eq. (9.1), a useful quantity is the receiver sensitivity, $P_r = NF + SINR + N0$.

Coverage can also be expressed by the maximum path loss (MPL) for a radio technology. Here path loss is the loss in the signal path due to distant propagation, building penetration, etc. of the radiated power. Hence, MPL can be calculated by the difference in radiated power levels at the transmitting and receiving antennas. To determine the MPL also the antenna gain at the transmitter and receiver need to be considered.

$$MPL\text{(dB)} = P_{T_x} + Gain_{T_x} + Gain_{R_x} - (NF + SINR + N0). \qquad (9.6)$$

An illustration of MCL and MPL design parameters is included in Fig. 9.3.

9.3 NB-IoT: PHY and MAC Characteristics

Unlike what happens with the use of ISM unlicensed bands, alternative cellular IoT solutions use GSM, WCDMA and/or LTE licensed bands. Despite its attractive and popular features, there are two basic drawbacks of ISM bands: they introduce time and power constraints. Specifically, based on the limited power of the IoT UL transmitter design and in an increasingly saturated ISM frequency space, short range IoT users represent a strong interference for long range IoT devices. In addition, the time constraints in the ISM band, such as the dwell time (time of permanence) or the duty cycle (work cycle), limit the coverage (it is a limit on the energy used) and/or the capacity of the system.

The main design premises of NB-IoT, related to other 3GPP IoT technologies like EC-GSM or LTE-M, are low device and deployment costs. Furthermore, NB-IoT is amenable to different licensed bands. It is compatible with the *frequencies re-farming* concept (re-use of old licensed bands) and also amenable with frequency bands used by 2G, 3G, and 4G cellular networks. 3GPP adopted a two-step strategy to cope with technological challenges brought by IoT services. The first step is a transition strategy that aims to utilize and optimize the existing network and technologies to provide IoT services [12]. The second step is a long term strategy based on the introduction of a new air interface technology for NB-IoT in order to support large-scale growth of MTC services and to maintain its core competitiveness toward non-3GPP LPWA technology [13].

NB-IoT reuses the LTE design extensively: numerology (frame design), OFDMA in the downlink, SC-FDMA in the uplink, simple turbo channel coding, etc. In addition to aiming low cost of devices and efficient power consumption, it is designed to have a reduced time to develop and also to have NB-IoT products for existing LTE equipment and software vendors.

NB-IoT supports three types of deployment scenes: (1) independent deployment (standalone mode), which utilizes independent frequency band that does not overlap with the frequency band of LTE; (2) guard-band deployment (guard-band mode), which utilizes edge frequency band of LTE; in-band deployment (in-band mode), which utilizes LTE frequency band, and it takes one PRB of LTE frequency band resource for deployment.

In the DL, OFDM is applied using a 15 kHz subcarrier spacing with normal cyclic prefix (CP). Each of the OFDM symbols consists of 12 subcarriers occupying this way the bandwidth of 180 kHz. Seven OFDMA symbols are bundled into one slot, so that the slot has the resource grid as illustrated in Fig. 9.4a. This is the same resource grid as for LTE in normal CP length for one resource block, which is important for the in-band operation mode. A resource element is defined as one subcarrier in one OFDMA symbol, represented by one square of Fig. 9.4a. Each of these resource elements carries a complex value with values according to the modulation scheme. These slots form subframes and radio frames in the same way as for LTE, as illustrated in Fig. 9.4b. There are 1024 cyclically repeated radio frames, each of 10 ms duration. A radio frame is partitioned into 10 subframes,

Fig. 9.4 NB-IoT physical resource block and frame structure

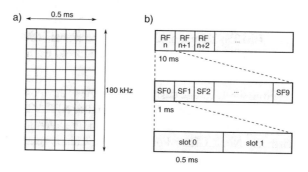

each one composed of two slots. In addition to the system frames, also the concept of hyperframes is defined, that spans 1024 system frame periods, corresponding to a time interval of almost 3 h.

9.3.1   Signals and Channels

Relevant physical signals and channels used in NB-IoT are the following (Fig. 9.5):

- Downlink:
  - **NPSS, NSSS, NRS**: Narrowband primary and secondary synchronization signals (NPSS and NSSS) are designed to allow a device to use a unified synchronization algorithm during initial acquisition. Narrowband reference signal (NRS) is used to allow the device to estimate the DL propagation channel coefficients and perform DL signal strength and quality measurements.
  - **NPBCH**: Narrowband physical broadcast channel (NPBCH) is used to deliver the NB-IoT Master Information Block (MIB), which provides essential information for the device to operate in the NB-IoT network (operation mode, frequency raster, scheduling, congestion control mechanisms, etc.).
  - **NPDCCH, NPDSCH**: Narrowband physical DL control and shared channels. NPDCCH is used to carry DL control information, in particular: (1) UL grant information, (2) DL scheduling information, (3) indicator of paging. NPDSCH is used to transmit unicast data and broadcast information such as system information (SI) messages.

- Uplink
  - **NPRACH**: Narrowband physical random access channel is used by the device to initialize connection and allows the serving base station to estimate the time of arrival (ToA) of the received NPRACH signal. Up to three NPRACH configurations can be used in a cell to support devices in different coverage classes.
  - **NPUSCH**: Narrowband physical uplink shared channels. The NPUSCH (in two formats) is used to carry UL user data and control information from higher layers. Additionally, NPUSCH also carries Hybrid Automatic Repeat Request (HARQ) acknowledgment for NPDSCH.
  - **DMRS**: Demodulation reference signal. DMRS is always associated with NPUSCH. It allows the base station to estimate the UL propagation conditions.

From the coverage design point of view, the most important channels to be taken into account are NPBCH in the DL and/or NPRACH in the UL. However, based on the perspective of energy consumption, the attention must be put on the dedicated channels NPDCCH and/or NPUSCH. This is related to their critical functions to

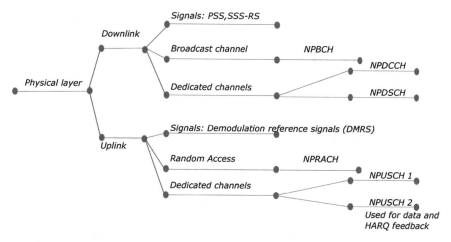

Fig. 9.5 NB-IoT frame signals and channels

begin specific tasks to obtain a radio resource control (RRC) connection as defined in the standard specification (Fig. 9.5).

9.3.2   State Model

Since UE and base station communication requires a radio resource control (RRC) connection to be established, it is important to pay attention to the key aspects of energy consumption [14] of that procedure. There are six basic states grouped according to the RRC available that define the way the UE NB-IoT works:

- Idle RRC state: Idle mode, Power saving mode, Paging mode.
- Connected RRC state: Connected, Uplink, Downlink.

The UE in RRC connected state consumes more energy (it has dedicated channels to begin the data transmission and needs to monitor the NPDCCH). NPDCCH is required to receive the DL data notification or UL data grant from the base station. When the UE releases its active RRC connection, it moves to the RRC Idle state. Related to the transition among these states, NB-IoT uses two energy saving mechanisms: power saving mode (PSM) and extended discontinuous reception (eDRX) (Fig. 9.6).

In PSM mode the UE is still registered to the network but cannot be reached by signaling in order to make the terminal deep sleep for a longer time to achieve the power saving. On the other hand, eDRX extends sleep cycle of UE in idle mode and reduces unnecessary startup of signals from the base station. These mechanisms modify the pattern of UE communication with the network using (four)

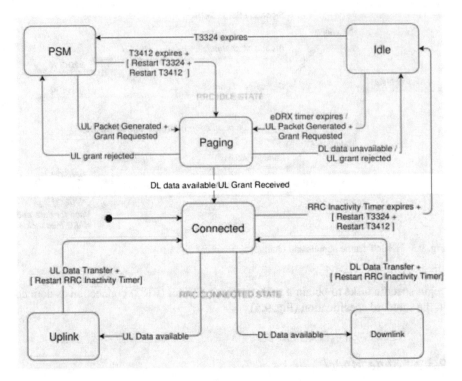

Fig. 9.6 State diagram of NB-IoT with emphasis in energy consumption [14]

RRC state transition timers. A brief description of each energy saving mechanism is the following:

- Extended Discontinuous Reception (eDRX): eDRX specifies the timers to deactivate monitoring of NPDCCH. eDRX works in cycles, where each cycle consists of a period during which the UE monitors the NPDCCH and a period during which the device saves its battery and stops monitoring the control channel (Fig. 9.7). The monitoring of the control channel for DL data indication or UL grant is known as Paging. During these periodic eDRX cycles, the UE is said to be in the Idle state. It remains in the Idle state until the expiration of a timer (called T3324) and then switches to another state called power saving mode (PSM). Since Paging consumes more energy that Idle state, they are illustrated in Fig. 9.6 as different states. If the monitored control messages contain a DL data indication or UL grant, the device switch its state to Connected. eDRX exists in both Connected and Idle RRC state, but is far more important in the Idle RRC state (due to its length) since it introduces more energy savings. As a consequence, eDRX is only considered in the Idle state in Fig. 9.6.
- Power Saving mode: PSM works in RRC Idle state enabling the device to enter into deep sleep. In the deep sleep state, the UE is unreachable by the network

Fig. 9.7 Connected eDRX: DRX cycles extended from 2.56 to 9.22 s. Idle eDRX: new paging time window allows long paging cycles (3 h)

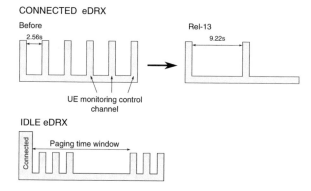

Fig. 9.8 T3324 timer determines for how long the UE will monitor paging before entering in PSM

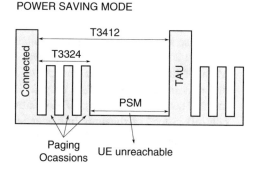

but stays registered to it (Fig. 9.8). The PSM cycle includes periodic reception of Paging messages and deep sleep. For most IoT scenarios, the downlink latency is not important and it is enough to monitor Paging infrequently which enables a device to be in deep sleep for a long time. For NB-IoT, a PSM cycle timer (called T3412 extended) is taking care of this function. If a UE in the PSM state generates a UL packet, it switches to the Paging state to monitor the control channel for the UL grant. The device will switch to Connected state if it receives the grant. If the grant is rejected, the device switches back to the PSM state. If there is a DL data packet for the device during the deep sleep state, the network buffers the packet and sends it when the device exits the PSM cycle. The device exits PSM on the expiration of timer T3412 extended to notify the network about its availability.

Compared to PSM, eDRX saves less energy but provides better downlink communication latency. Therefore, both the PSM and eDRX mechanisms can be used to adapt to different IoT scenarios.

The state diagram represents two more states in RRC Connected state, named Downlink and Uplink. In the Downlink state, the UE is awake and the base station transmits DL packets to the device. This state ends as soon as the base station has no DL packets for this device and so the ending time of this state is dynamic depending on the packet size and number of packets. Similarly, during the Uplink state, the UE transmits UL packets to the base station. When the device receives an RRC release message, it switches to the Idle state and restarts the timers T3324 and T3412.

9.3.3   Coverage, Efficiency, Capacity, Cost

We describe in the following some key characteristics of NB-IoT that cope with the objectives of extended coverage, energy efficiency, massive number of users and low cost.

- **Coverage**
  Based on [15] NB-IoT can reach 164 dB in independent deployment mode. Following the MCL discussion in Sect. 9.2.2, output power, signal bandwidth, and noise figure are not quantities suitable to obtain this coverage. Hence, only the SNR is the quantity left to improve.
  Most of the target service context of NB-IoT are mini-packet transmissions and it is hard for NB-IoT to provide long term and continuous indication of channel quality change [16]. NB-IoT introduces coverage level (open-loop power control) instead of dynamic link adaptation scheme [17]. There are three kinds of coverage classes including normal coverage, robust coverage and extreme coverage which correspond to MCL of 144 dB, 158 dB and 164 dB, respectively. For NPUSCH and/or NPRACH, the transmit power is the maximum configured device power, P_{max}. The maximum configured device power is set by the serving cell. Because NB-IoT maximum device power is 23 dBm, $P_{max} \leq 23$ dBm. The transmit power is determined by

$$P_{CH} = \max\{P_{max};\, 10\log(M) + P_t + \alpha L\} \quad \text{(dBm)}, \tag{9.7}$$

  where P_t is the target received power at the base station, L is the estimated path loss, α is a path loss adjustment factor, and M is a parameter related to the bandwidth of NPUSCH or NPRACH waveform. The bandwidth-related adjustment is used to relate the target received power to target received SNR. The device uses M according to NPUSCH or NPRACH transmission configuration. P_{max}, P_t, and α are provided by higher-layer configuration signaling.
  NB-IoT adopts data retransmission mechanism to obtain improved SNR. Also, time diversity gain and low-order modulation are used to improve demodulation and coverage performance [18, 19]. In case of limited coverage, the receiver will not be able to receive the signal of interest. This problem can be solved by the introduction of blind repetitions where the receiver first can accumulate the IQ representation of blindly transmitted bursts to a single burst with increased SNR before synchronizing to the burst and performing channel estimation. To maximize SNR, the signals should be coherently combined, i.e., aligned in amplitude and phase. Since the white noise (assumed uncorrelated) will be

non-coherently combined, that results in a lower noise power than in the wanted signal. Theoretically, when combining N transmissions, the SNR obtained SNR_N compared to a single transmission SNR_1 can be expressed as:

$$SNR_N \text{ (dB)} = 10\log(N) + SNR_1. \qquad (9.8)$$

However, a combination of two signals is rarely perfect, and signal impairments will result in an overall processing gain lower than in (9.8). To perform the coherent combination, the blind repetitions should also be coherently transmitted, i.e., multiple signals from the transmitting antenna follow the same phase trajectory. Without other impairments and considering a stationary propagation channel, the receiver will be able to accumulate the signal and obtain the SNR improvement.

Coverage extension in NB-IoT also relies on an improved channel coding for control channels, and efficient HARQ retransmissions for the data channels. The cost to obtain the specified MCL with a reliable data transmission is the increase of the latency (due to data retransmission). Currently, the tolerable latency in 3GPP NB-IoT is 10 s. In fact, lower latency of about 6 s for maximal coupling losses can be also supported [15].

- **Energy efficiency**
  By means of power saving mode and expanded discontinuous reception longer standby time can be realized in NB-IoT. NB-IoT requires that the UE service life of a constant volume battery is 10 years for typical low-rate low-frequency service. According to [15], for coupling loss of 164 dB and using both PSM and eDRX, the service life of 5-Wh battery can be 12.8 years if a message of 200 byte is sent once per day by the terminal.

  NB-IoT Release 13 supports two device power classes, 20 and 23 dBm. 23 dBm is the most common power class in LTE. The 20 dBm power class was specified with the target to simplify the integration of the PA on the chip to support a low-complexity system on chip design. The PA current expected for 20 and 23 dBm two power classes limits the ability of the so-called small coin cell batteries. To facilitate the use of coin cells to support NB-IoT operation, a 14 dBm power class is introduced in 3GPP Release 14. The UL coverage is negatively impacted by the reduction in device output power. The supported MCL for a 14 dBm device is reduced from 164 to 155 dB.

- **Number of connections (capacity)**
  As discussed before [20], the development of NB-IoT is based on LTE. The modification is mainly made on relevant technologies of LTE according to NB-IoT unique features. The RF bandwidth of NB-IoT physical layer is 200 kHz. In downlink, NB-IoT adopts QPSK and OFDMA modulation with subcarrier spacing of 15 kHz [21]. In uplink, BPSK or QPSK and SC-FDMA modulation are adopted. A single subcarrier technology with subcarrier spacing of 3.75 kHz and 15 kHz is addressed to user terminal with ultra-low rate and ultra-low power consumption.

To support a massive number of devices, NB-IoT also includes a multicarrier feature. In addition to the anchor carrier, which carries synchronization and broadcast channels, one or more non-anchor carriers (that does not carry NPBCH, NPSS, NSSS) can be provided. When a UE in idle mode selects an anchor carrier, it senses only the paging messages on the anchor. When the UE needs to switch from the idle to connected mode, the procedure that uses the NPRACH also takes place on the anchor carrier. The network can use the RRC configuration to point the UE to a non-anchor carrier. Essential information about the non-anchor carrier is provided using dedicated signaling. During the remaining duration of the connected mode, user monitor NPDSCH and/or NPUSCH activities that take place on the assigned non-anchor carrier. After the device completes the data session, it comes back to select the anchor carrier during the idle mode.

The non-anchor carriers can be allocated according to the traffic load of NB-IoT. Many IoT use cases generate highly delay-tolerant traffic. Such traffic can be delivered to the network during non-critical hours. The multicarrier feature allows the radio resources normally reserved for serving broadband or voice to be allocated for NB-IoT when the load of broadband and voice traffic is low. The network can, as an example, take advantage of the multicarrier feature to push firmware upgrades to a massive number of devices during the middle of the night.

The basic design of NB-IoT (3GPP Rel. 13) aimed at supporting at least 60,680 devices/km^2. In 3GPP Release 14, the target was increased to support 1,000,000 devices/km^2 as a consequence of a desire to make NB-IoT a 5G technology [22]. That translates to approximately 112 access arrivals per second. To keep the collision rate of competing preambles low, a sufficiently high amount of radio resources need to be reserved for the NPRACH. This, in combination with the required support of devices in extended coverage, motivated the introduction of random access on the non-anchor carriers. Also, to support UE accessibility for a very high number of users, paging support has been added to the non-anchor carriers in 3GPP Release 14.

- **Cost**

An NB-IoT module can be implemented as a system on chip (SoC). The functionality on the SoC can be typically divided into the following main components : peripherals, real time clock (RTC), central processing unit (CPU), digital signal processor (DSP) and hardware accelerators, and the radio transceiver (TRX). Also, certain parts may be located outside the SoC, as for example: the power amplifier (PA) and crystal oscillators (XO).

The NB-IoT module complexity and cost are mainly related to the complexity of baseband processing, memory consumption, and radio-frequency (RF) requirements. Regarding baseband processing, NB-IoT is designed for allowing low-complexity receiver processing during initial cell selection and during connection.

For initial cell selection, a device needs to search for only one synchronization sequence for establishing basic time and frequency synchronization to the network. The device can use a low sampling rate (e.g., 240 kHz) and take advantage of the synchronization sequence properties to minimize memory and complexity. During connected mode, low device complexity is facilitated by restricting the DL transport block size and relaxing the processing time requirements. For channel coding, NB-IoT adopts a simple convolutional code [23] in the DL channels. In addition, a device needs to support only half-duplex operation (a duplexer is not needed in the RF front-end).

Regarding RF, all the performance objectives of NB-IoT can be achieved with one transmit-and-receive antenna in the device. NB-IoT is designed for allowing relaxed oscillator accuracy in the device. For example, a device can achieve initial acquisition when its oscillator inaccuracy is as large as 20 parts per million (ppm). During a data session, the transmission scheme is designed for the device to easily track its frequency drift.

In NB-IoT, in principle only a 16-point FFT is needed. The complexity of N-point FFT is $6N \log_2 N$ real operations. There are 14 OFDM symbols per subframe, and therefore the complexity associated with FFT demodulation is approximately 5376 real-value operations per subframe. The complexity of decoding a block size of 680 bits is around 260 k operations. Considering the most computationally demanding case in Release 13 (receiving 3 block size and have 12 ms before signaling HARQ-ACK), the complexity of FFT and decoding is approximately 23 millions of operations per second (MOPS). Regarding cell selection or reselection procedures, the complexity is dominated by NPSS detection [24] (required to calculate a correlation value per sampling time interval), that is less than 30 MOPS. As a consequence, an NB-IoT device can be implemented with baseband complexity less than 30 MOPS.

9.4 LTE and NB-IoT Coexistence: Interference Due to RF Impairments

NB-IoT considers the coexistence of IoT devices with LTE UE, by using the same basic unit of resource allocation, i.e., both systems use the same PRB [25, 26]. Among the operation modes defined, we focus in this analysis in the in-band mode.

Considering the 15 kHz subcarrier spacing, the PRB occupies a bandwidth of 180 kHz. In conventional applications (LTE-A), multiple PRB are allocated to a single user. However, in case NB-IoT nodes, the use of sub-PRB transmission is also considered. For DL, NB-IoT uses the same intercarrier spacing than LTE. On the other hand, in UL a single tone mode with an intercarrier spacing of 3.75 kHz is included [27]. This mode is used by low data rate power-constrained nodes, since the single carrier modulation leads to an efficient RF implementation, due to the low peak-to-average power ratio transmission. Additionally, the reduced channel bandwidth requires a quarter of the sampling frequency, and it allows the allocation of 48 UE in the same PRB [28].

In the base station, it is possible to use a common analog front-end to demodulate LTE and single tone NB-IoT signals. However, the difference in the PRB structure destroys the orthogonality and creates interference. This situation affects the coexistence of LTE and NB-IoT users and needs to be carefully studied. Moreover, the low cost constraint in NB-IoT devices, creates several RF impairments that degrade the system performance. The coexistence between LTE and NB-IoT devices is studied considering the interference produced not only by different structure of LTE and NB-IoT PRBs but also by the most harmful RF impairments.

This section shows the results of the analysis of the bandwidth required to obtain adequate isolation between the LTE and NB-IoT systems [29]. Additionally, we show the allowable level of CFO and I/Q imbalance that the IoT uplink can tolerate without compromising the performance.

9.4.1   Modeling Signals and Interference

The scenario is a base station serving several LTE and NB-IoT UEs at the same time. The base station allocates PRBs to different LTE or NB-IoT devices according to the network load. For the downlink, NB-IoT nodes use LTE numerology, i.e., $1/T_L =$ 15 kHz subcarrier spacing and a subframe of 1 ms formed by two slots of 7 OFDM symbols. An LTE PRB is formed by $M_L = 12$ contiguous subcarriers, during a slot. As in the OFDMA case, there is no intercarrier or interuser interference in the downlink given that the base station is a time and frequency reference for different UE.

On the other hand, there are three possibilities for the NB-IoT uplink: multitone transmission based on SC-FDMA with 15 kHz of intercarrier spacing, single-tone also with 15 kHz, and single tone with $1/T_I = 3.75$ kHz. In the 3.75 kHz single tone mode, the PRB M_I has 48 subcarriers (or different UEs) with a slot of 2 ms. A diagram for the coexistence of LTE and NB-IoT PRB is depicted in Fig. 9.9. This case is of particular interest since it allows a significant reduction in the complexity of the UE transmitter implementation, and increases the amount of UEs simultaneously connected to the BS. It is important to note that due to the different

Fig. 9.9 PRB structure for the coexistence of LTE and IoT in the Uplink

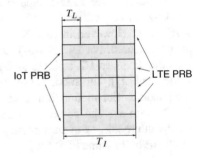

numerology, there is interference between LTE and UEs, i.e., this scenario is not equivalent to an OFDMA system.

First, the complex base-band symbols are extended to consider the cyclic prefix (CP) N_{cpI} and then are converted to the analog domain. After that, they are modulated in a carrier of frequency $f_c + (k + iM_I)/T_I$, corresponding to the subcarrier k of the i-th PRB, centered at f_c. Without loss of generality, UE k is the node allocated to subcarrier k of PRB i. The discrete baseband equivalent of the NB-IoT UE k transmitted signal is

$$x_{k,i,\ell}(n) = X_{k,i,\ell} \exp\left(\frac{j2\pi n}{N_I}(k + iM_I)\right) \tag{9.9}$$

for $-N_{cpI} \le n \le N_I$, where $X_{k,i,\ell}$ is the Q-PSK symbol sent by the user k in the block ℓ, N_I is the amount of samples in the NB-IoT symbol, and N_{cpI} the CP.

The LTE signal represents an interference to the PRB assigned to NB-IoT UE. The transmitted signal by the LTE UE results

$$v_\ell(n) = \sum_{k=0, k\notin \mathscr{I}_{IoI}}^{N_L-1} V_{k,\ell} \exp\left(\frac{j2\pi kn}{N_L}\right) \text{ for } - N_{cpL} \le n \le N_L, \tag{9.10}$$

where $V_{k,\ell}$ are M-QAM symbols sent by LTE UEs, $\mathscr{I}_{IoI}$ is the index set of the carriers that are allocated to NB-IoT UEs, $N_L = N_I/4$ is the OFDM symbol length, and N_{cpL} the CP. Note that different LTE UEs are not differentiated since the analysis is concentrated in overall interference they generate.

NB-IoT channel of user k of PRB i is defined as $h_{k,i}(n)$, and the LTE channel is defined as $h_L(n)$. Assuming the CPs length of LTE and NB-IoT modulations longer than the channel impulse response, then the received signal at the BS can be expressed as:

$$y_\ell(n) = \sum_{i=0, i\in \mathscr{R}_{IoI}}^{N_{rb}-1} \sum_{k=0}^{M_I-1} H_{k,i}(k + iM_I)X_{k,i,\ell} \exp\left(\frac{j2\pi n}{N_I}(k + iM_I)\right)$$
$$+ \sum_{k=0, k\notin \mathscr{I}_{IoI}}^{N_L-1} H_L(k) V_{k,\ell} + \exp\left(\frac{j2\pi kn}{N_L}\right) + w(n), \tag{9.11}$$

where $\mathscr{R}_{IoI}$ is the index set of PRB allocated to NB-IoT, $N_{rb} = N_L/M_L = N_I/M_I$ is the total amount of PRBs, $H_{k,i}(k)$ is the N_I-length discrete Fourier transforms (DFT) of $h_{k,i}(n)$, $H_L(k)$ is the N_L-length DFT of $h_L(n)$, and $w(n)$ is AWGN.

As the sample rate of both systems is the same ($T_L/N_L = T_I/N_I$), the receiver at the BS can be implemented with a common RF front-end. After the down-conversion, the samples are rearranged to form subframes in each system, according to the amount of samples defined previously (N_{cpL}, N_{cpI}, N_L, and N_I). It is important to note from (9.11) that the demodulation of NB-IoT is performed by an N_I-FFT, whereas the LTE demodulation by an N_L-FFT. Since the PRBs have different length, there is interference between the systems.

Next, possible interference scenarios are considered to study how LTE UEs and hardware imperfections affect the performance of NB-IoT devices. NB-IoT nodes are low power and low cost devices. Since power constraint limits the transmitted power, the interference of NB-IoT over the LTE signal can be neglected. The focus in this analysis is on the interference of LTE over NB-IoT UEs, since it is of more practical importance. On the other hand, low cost devices are prone to have large RF impairments and synchronization offsets that produce intercarrier and interuser interference, that reduce the system performance. Additionally, both constraints imply that the computation capacity is limited.

As discussed before, the LTE signal is not orthogonal to the NB-IoT. The power leakages from LTE PRBs to those of NB-IoT and has a *sinc* envelope given by the squared symbol pulse shape in time domain. Then, a simple method to reduce the interference between the systems is to include guard bands between PRBs belonging to LTE and IoT.

In order to allow a large number of simultaneous transmitting UEs, NB-IoT considers a mode with 3.75 kHz of intercarrier spacing. Although this also reduces the implementation complexity, it makes the system more sensitive to RF imperfections related to the oscillator and mixer inaccuracies, such as carrier frequency offset, I/Q imbalance, and phase noise. For frequencies around 2 GHz, as the used in LTE-A networks, oscillators are quite stable and the phase noise does not pose a problem. It is assumed that the time offset is corrected in the acquisition phase. For this reason, it is considered only CFO and I/Q imbalance in the analysis. Different to the interference between LTE and NB-IoT, the CFO and I/Q imbalance cause interference between users that belong to the same PRB.

9.4.2 Numerical Results

In this section there are illustrated by simulations the interference scenarios discussed. First, it is shown the effectiveness of guard bands in NB-IoT PRB to mitigate the interference of the LTE signal. Then, we analyze the intra-PRB interference of NB-IoT UEs due to CFO and I/Q imbalance.

It is considered that the UL where LTE and NB-IoT UEs coexist. System parameters are defined in Table 9.3. Each carrier in the NB-IoT PRB belong to a different user with its own hardware and synchronization imperfections. In the simulation it is considered only an NB-IoT PRB, since it is the worst interference case.

In Fig. 9.10, it is depicted the averaged BER of NB-IoT UEs that belong to the PRB, under the interference of LTE UEs that fully complete the rest of the PRBs. The figure illustrates the effect of different guard bands in the mitigation of the LTE interference. The guard bands are allocated at the edge of the NB-IoT PRB. Both signals have the same received power. It is evident that even large guard bands do

Table 9.3 Summary of system parameters

Parameter	LTE	IoT
Subcarriers	$N_L = 1024$	$N_I = 4096$
CP	$N_{cpL} = 16$	$N_{cpI} = 64$
PRB size	$M_L = 12$	$M_I = 48$
Inter. spacing	$1/T_L = 15\,\text{kHz}$	$1/T_I = 3.75\,\text{kHz}$
Constellation	64-QAM	Q-PSK

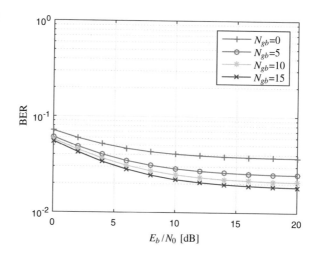

Fig. 9.10 Averaged BER of NB-IoT UEs under LTE interference. Effect of guard bands at the edge of the NB-IoT PRB

not allow to obtain reasonable BER performance. This is a consequence of the slow decay of the *sinc* function. It can be concluded that a more sophisticated technique is needed to compensate for the LTE interference, in order to get a better performance without sacrificing bandwidth.

To evaluate the BER degradation due to the CFO, it is considered a single NB-IoT UE without active LTE UEs. In Fig. 9.11, it is depicted the average BER of NB-IoT users for different CFO ranges. The CFO is normalized to the intercarrier spacing, different for each user, and uniformly distributed in the range $\{-\epsilon_L, \epsilon_L\}$. Following the figure it can be noted that the average BER of UEs in the NB-IoT PRB is adequate for CFO values below 10%. This only can be achieved for static applications and if the NB-IoT UE synchronizes with the BS in the downlink, prior to the uplink transmission.

Finally, the degradation in the averaged BER due to I/Q imbalance is shown in Fig. 9.12. The parameters α and θ are respectively the amplitude and phase mismatches of the mixer. As can be concluded from the figure, the I/Q imbalance does not cause a noticeable performance drop due to the low constellation used by NB-IoT devices.

Fig. 9.11 Averaged BER of the IoT UEs due to CFO. No active LTE users

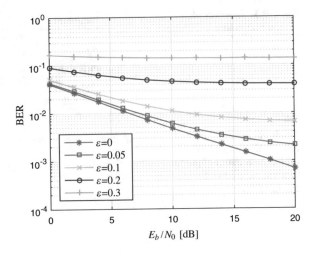

Fig. 9.12 Averaged BER of the IoT UEs due to I/Q imbalance. No active LTE users

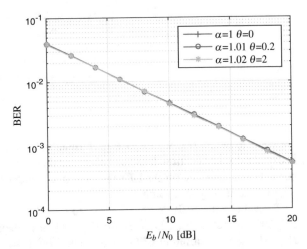

9.5 Summary of the Key Points

Despite the different design starting point of some proprietary IoT technologies, addressed to optimize consumption, cellular based NB-IoT has proven to be very competitive in that aspect. As a consequence, other characteristics must be taken into account when adopting the IoT technology for an specific application. For example [30]:

- Interferences: NB-IoT available multiple access techniques are efficient and very well known. However, even for other IoT technologies, the use of non-orthogonal multiple access (NOMA) could be an interesting alternative to obtain good results in extreme coverage [31].

- Spectrum: NB-IoT coexists with LTE in a proprietary part of the spectrum. Technologies using ISM bands share the spectrum and may be subject to external interference and/or may suffer of capacity constraints.
- QoS: The NB-IoT network guarantees delivery. This is an important aspect because alternatives like LoRa can incur significant energy costs for guaranteed delivery, as they are also severely limited by duty-cycle regulations.
- Latency: The price to pay for low consumption in NB-IoT is high variability in delivery time.
- Data rate: Different IoT technologies have been designed to transmit a few bytes per hour, even per day. If the application sporadically requires high bandwidth, NB-IoT may be a good option.
- Network model: NB-IoT is offered as a connectivity service under a contract that charges a set price for each transmitted byte. The infrastructure is owned by an operator and hence signal coverage depends on the deployed infrastructure. That limits the possibilities of the application [32]. As an alternative, LoRa allows the user to reduce the energy consumption of the devices by deploying a closer gateway.

Concluding the chapter we analyze the feasibility of the uplink coexistence of LTE and NB-IoT UE. We consider the NB-IoT single tone protocol with 3.75 kHz intercarrier spacing. The coexistence with LTE introduces interference in the NB-IoT receiver, and the reduction of the intercarrier spacing makes the system less robust to carrier frequency offset (CFO) and the I/Q imbalance. The addition of guard bands at the edges of the NB-IoT PRB does not provide the necessary isolation level and the performance of the system is compromised. On the other hand, we show that the NB-IoT receiver is robust to CFO levels below 10% and that the I/Q imbalance does not produce a noticeable drop in the performance.

The aim of this chapter is just to introduce design parameters of some key IoT technologies, with emphasis in NB-IoT considering the evolution and results obtained in the cellular based communication industry. However, there is no specific conclusion in favor of any IoT solution, except the careful analysis of the design aspects of the application at hand.

References

1. Q. Qi, F. Tao, Digital twin and big data towards smart manufacturing and industry 4.0: 360 degree comparison. IEEE Access **6**, 3585–3593 (2018)
2. H. Sun, Ch. Wang, B. Ahmad, *From Internet of Things to Smart Cities: Enabling Technologies* (CRC Press, Boca Raton, 2017)
3. Sigfox, Technology overview. www.sigfox.com/en/sigfox-iot-technology-overview (2019)
4. Semtech, LoRa. www.semtech.com/lora (2019)
5. LoRa Alliance Inc., Lorawan specification. V1.0.2 (2016)
6. Ingenu, How RPMA works, White paper (2017)
7. Samsung Semiconductor, IoT Exynos i S111 - NB-IoT solution to make connection go further. www.samsung.com/semiconductor/minisite/exynos/products/iot/exynos-i-s111/ (2019)

8. 3rd Generation Partnership Project (3GPP), Technical specification group radio access network; evolved universal terrestrial radio access (E-UTRA); LTE/WLAN radio level integration using IPsec tunnel (LWIP) encapsulation; protocol specification (release 15). 3GPP TS 36.361 v15.0.0 (2018)
9. Y. Roth, The physical layer for low power wide area networks: a study of combined modulation and coding associated with an iterative receiver, PhD Dissertation, University of Grenoble Alpes (2017)
10. MulteFire Alliance, MulteFire specifications. MulteFire Release 1.1 Enhancements (2019)
11. E. Jorgensen, Cell acquisition and synchronization for unlicensed NB-IoT. Ph.D. Thesis, Department of Electrical Engineering, Linköping University (2017)
12. Ericsson, Cellular networks for massive IoT. Technical Report, Stockholm (2016)
13. RIoT, Low power networks hold the key to Internet of Things. Technical Report, Berlin (2015)
14. A.K. Sultania, P. Zand, C. Blondia, J. Famaey, Energy modeling and evaluation of NB-IoT with PSM and eDRX, in *2018 IEEE Globecom Workshops (GC Wkshps)* (2018), pp. 1–7
15. D. Guo-Hua, Y. Jun-Hua, Research on NB-IoT background, standard development, characteristics and the service. IEEE Trans. Veh. Technol. **40**(7), 31–36 (2016)
16. J.J. Nielsen, D.M. Kim, G.C. Madueno, N.K. Pratas, P. Popovski, A tractable model of the LTE access reservation procedure for machine-type communications, in *2015 IEEE Global Communications Conference (GLOBECOM)* (2015), pp. 1–6
17. M.T. Islam, A.M. Taha, S. Akl, A survey of access management techniques in machine type communications. IEEE Commun. Mag. **52**(4), 74–81 (2014)
18. F.A. Tobagi, Distributions of packet delay and interdeparture time in slotted aloha and carrier sense multiple access. J. ACM **29**(4), 907–927 (1982)
19. J. Seo, V.C.M. Leung, Design and analysis of backoff algorithms for random access channels in UMTS-LTE and IEEE 802.16 systems. IEEE Trans. Veh. Technol. **60**(8), 3975–3989 (2011)
20. Z. Yulong, D. Xiaojin, W. Quanquan, Key technologies and application prospect for NB-IoT. ZTE Technol. **23**(1), 43–46 (2017)
21. X. Ge, X. Huang, Y. Wang, M. Chen, Q. Li, T. Han, C. Wang, Energy-efficiency optimization for MIMO-OFDM mobile multimedia communication systems with QoS constraints. IEEE Trans. Veh. Technol. **63**(5), 2127–2138 (2014)
22. 3rd Generation Partnership Project (3GPP), Technical specification group radio access network; physical layer aspects for evolved universal terrestrial radio access (UTRA); radio resource control (RRC); protocol specification, 3GPP TR 36.331 v13.3.0 (2016)
23. 3rd Generation Partnership Project (3GPP), Technical specification group radio access network; physical layer aspects for evolved universal terrestrial radio access (UTRA); multiplexing and channel coding. 3GPP TR 36.331 v13.3.0 (2016)
24. Qualcomm, 3GPP technical specification group radio access network; physical layer aspects for evolved universal terrestrial radio access (UTRA); r1-161981 NB-PSS and NB-SSS design. 3GPP TSG RAN1 Meeting 2 NB-IoT (2016)
25. H. Malik, H. Pervaiz, M. Mahtab Alam, Y. Le Moullec, A. Kuusik, M. Ali Imran, Radio resource management scheme in NB-IoT systems. IEEE Access **6**, 15051–15064 (2018)
26. A. Ijaz, L. Zhang, M. Grau, A. Mohamed, S. Vural, A.U. Quddus, M.A. Imran, C.H. Foh, R. Tafazolli, Enabling massive IoT in 5G and beyond systems: PHY radio frame design considerations. IEEE Access **4**, 3322–3339 (2016)
27. 3rd Generation Partnership Project (3GPP), Technical specification group radio access network; physical layer aspects for evolved universal terrestrial radio access (UTRA); NB-IoT; technical report for (BS) and (UE) radio transmission and reception (release 13). 3GPP TR 36.802 V13.0.0 (2016)
28. L. Zhang, A. Ijaz, P. Xiao, R. Tafazolli, Channel equalization and interference analysis for uplink narrowband internet of things (NB-IoT). IEEE Commun. Lett. **21**(10), 2206–2209 (2017)
29. G.J. Gonzalez, F.H. Gregorio, J. Cousseau, Interference analysis in the LTE and NB-IoT uplink multiple access with RF impairments, in *2018 IEEE 23rd International Conference on Digital Signal Processing (DSP)* (2018), pp. 1–4

30. B. Martinez, F. Adelantado, A. Bartoli, X. Vilajosana, Exploring the performance boundaries of NB-IoT. IEEE Internet Things J. **6**(3) 5702–5712 (2019)
31. M. Bernhardt, J.E. Cousseau, On interference alignment based NOMA for downlink multicell transmissions, in *2018 IEEE 23rd International Conference on Digital Signal Processing (DSP)* (2018), pp. 1–5
32. O. Liberg, M. Sundberg, Y. Wang, J. Bergman, J. Sachs, *Cellular Internet of Things: Technologies, Standards, and Performance* (Academic Press, London, 2018)

Chapter 10
Final Notes and Novel Issues

Abstract This chapter summarizes the most relevant topics addressed in the book, including novel interference cancellation techniques, implementation bottlenecks, and open issues that will require future research. The state-of-the-art techniques for increasing energy and spectral efficiency in 5G cellular system are also briefly described.

10.1 5G Implementation Challenges

Insights on advanced topics regarding the implementations of the state-of-the-art technologies for novel communication systems focused on 5G are presented in this chapter. This book addressed a comprehensive treatment of the ongoing research into the novel technologies to meet the demands of high spectral/energy efficiency and higher aggregate throughput imposed by 5G. The implementation of full-duplex systems to reach high spectral efficiency is treated in Chap. 7. In addition, massive MIMO implementations are described in Chap. 8. Energy efficiency and implementation issues are also addressed in that chapter.

Open research topics include the combination of massive MIMO and full-duplex (FD), the use of millimeter-wave frequency bands for massive MIMO systems, and the implementation challenges imposed by MaMIMO. These future research topics are discussed in the following sections.

10.1.1 The Combination of Massive MIMO and Full-Duplex

One of the legs of the so-called *5G triangle*, depicted in Fig. 10.1, are the *enhanced broadband* communications (EMBB). To achieve these throughput figures while maintaining high spectral and energy efficiencies, full-duplex operation mode and MaMIMO systems were recently introduced. For a detailed revision of those techniques, see Chaps. 7 and 8, respectively. The development of FD technology

© Springer Nature Switzerland AG 2020

F. Gregorio et al., *Signal Processing Techniques for Power Efficient Wireless Communication Systems*, Signals and Communication Technology,
https://doi.org/10.1007/978-3-030-32437-7_10

Fig. 10.1 5G triangle

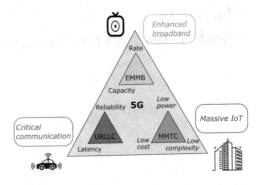

allows for co-channel simultaneous transmissions, but suffers from an unavoidable self-interference (SI). On the other hand, massive MIMO (MaMIMO) considers transceivers with a huge amount of transmitting and receiving antennas that allows for spatial-multiplexing and interference/noise mitigation techniques. Therefore, it results natural to complement both techniques to enhance the system performance. Base stations and relays, used in cellular heterogeneous networks, can benefit from the combined use of FD and MaMIMO techniques.

In a cellular network composed of FD base stations and FD users, in addition to the SI, there is a performance drop due to the interference between the downlink (DL) and uplink (UL). This interference increases with the nodes density. In this context, it is known that FD improves significantly the spectral efficiency of DL, while UL remains as a bottleneck for the overall system performance. This bottleneck is caused by the interference between DL and UL (due to the large disparity between base station and users transmitted powers) as well as the imperfect SI cancellation. The additional degrees of freedom introduced by MaMIMO can be used to mitigate that interference and the residual SI, and thus improve the spectral efficiency of the UL. These achievements are conditioned to the use of effective SI cancellation techniques in the user devices. If that is not the case, FD technology should be used only in the base station [1].

The deployment of many small cell base stations (SBS) can be used to reduce the inter-cell interference and improve the performance. As a consequence, the problem that arises is how to forward the backhaul traffic to the core network. The excessive cost and lack of flexibility of wired connections and the poor penetration and installation complexity of dedicated millimeter wave links motivate the use of self-backhaul SBS. These solutions use MaMIMO backhaul hubs to simultaneously handle the access and the backhaul links in the same radio channel. On the other hand, hybrid HD/FD SBS improves the utilization of the radio channel. The probability of achieving a certain rate in the downlink decreases as the proportion of SBS using HD increases. The same result is verified for the FD mode. This is due to the increment of intra-cell interference as the number of interfering SBS increases. Therefore, there is a proportion of SBS using HD and FD modes that maximizes the system performance [2].

In MaMIMO systems, the correlation among antennas affects in general the performance. For the particular case of FD two-way relays, the asymptotic sum-rate is independent of the antenna correlation at the transmitter side of the relay. On the other hand, it is highly affected by the antenna correlation at the source and destination (and in a lesser extent by the correlation at the relay receiver). In other words, a high correlation in those antennas results in a lower sum-rate. Additionally, in order to achieve a non-vanishing sum-rate as the number of antennas increases, the power scaling at the source and the relay should be $P_S = E_S/M_R^k$ and $P_R = E_R/M_R^q$ with $k, q \geq 0$, where E_S and E_R are the power budget, and M_S and M_R are the number of antennas, respectively at the source and relay [3].

From the previous discussion is undoubtedly that the combination of FD and MaMIMO boost the achievable sum-rate. However, this gain comes at the price of a considerable growth in the hardware complexity associated with the huge amount of implemented antennas. As discussed in Chaps. 5 and 7, communication systems need fast high-resolution ADCs to achieve the stringent low-distortion figures and handle the increasing operating bandwidth. It is well known that the ADC power consumption scales exponentially with the resolution and linearly with the sampling rate. Additionally, the chip area also increases exponentially with resolution, increasing costs and limiting the integration of several front-ends. This poses a challenge in MaMIMO systems since two ADCs are required for each antenna front-end.

A possible solution to this problem is to reduce the resolution of ADCs and use the MaMIMO additional degrees of freedom to compensate for the introduced distortion. However, the number of necessary antennas could be prohibitive. A one-bit ADC needs to double the amount of antennas in order to reach the sum-rate of a system with perfect (infinite resolution) ADCs. On the other hand, it has been shown that ADCs of 4 or 5 bits are optimal in terms of energy efficiency. For a FD MaMIMO relay, the resolution of the receiver ADCs is the bottleneck of the system performance, since it cannot be compensated by increasing the amount of antennas. Therefore, it is convenient to use higher resolution ADCs at the receiver than in the relay. This can be explained by the double quantization that the signal suffers at the relay and the receiver. Despite the low-resolution limiting factor, the degrees of freedom of MaMIMO allow to achieve a constant rate, with a fixed power budget, while scaling the transmitted power per antenna inversely proportional to the number of radiating elements [4].

To summarize the section, some open research opportunities can be envisioned. The flexibility of this system allows the implementation of new access protocols, considering heterogeneous multi-tier topologies and dual (HD/FD) operation modes. This requires the design of novel strategies that take advantage of the available degrees of freedom. In terms of physical implementation, the inclusion of a huge number of antennas is a major challenge. Simple and cheap but yet low-distortion implementations are required.

10.1.2   Millimeter Wave Wireless Communications

Massive MIMO beamforming (BF) techniques at higher frequency bands in millimeter waves between 24 and 52 GHz are considered a key technology to meet the greedy specifications of data throughput, channel capacity, and low latency in 5G new radio (NR) systems [5]. These technologies would allow the efficient use of new large portions of available spectrum [6]. In addition, they would also be practical tools to combat (or alleviate) the severe channel path-loss effects present at mm-waves [7]. By providing the base station with large number of antennas (in the order of $N \cong 100$), narrow beams with high directivity can be generated and oriented towards user equipment (UE). As a consequence, large increments on signal-to-interference-plus-noise ratio (SINR) can be spatially obtained by maximizing the energy radiated in the UE direction while minimizing interference radiation from other users [8].

Several general aspects of massive MIMO BF design for millimeter waves communications are described in [9] including: channel models, channel estimation techniques, architectures, pre-coding, and combining. Furthermore, a survey about different techniques and proposals to tackle associated challenges can be found in [10]. Beamforming techniques can be classified according to the domain in which the processing is performed before the signals are fed to the antennas. That means digital beamforming (DBF), analog beamforming (ABF), and hybrid analog-digital beamforming (HBF). A brief review on ABF and DBF is included in [11], where an analysis on achievable throughput and energy efficiency is presented, making focus on ABF and singular value decomposition (SVD) based DBF techniques.

Whereas in ABF only one RF chain is connected to the N_T transmission antennas, and the BF operation is therefore performed in the analog domain through analog phase shifting, the number of RF chains and antennas is equal in DBF in a one-to-one connection where the phase shifting and weighing operations are conducted in the digital domain. In HBF, each RF chain is connected to several antennas and two architectures exist, i.e., fully connected and partially connected. Each RF chain is connected to all the available antennas in fully connected HBF, while in the partially connected scenario each RF chain is connected to a subset of non-shared antennas [7, 11, 12].

According to [11], HBF is based on approximating SVD based DBF with a reduced number of radio frequency (RF) chains. This is due to the fact that, although it is assumed that it can offer the best performance in SINR, power consumption remains a concern regarding DBF since an RF chain is needed for each antenna. As a consequence, much research has been recently conducted in HBF techniques amenable to 5G NR [7, 12, 13]. In [7], for example, SVD based DBF is combined with a constant modulus algorithm for ABF in which the power of the desired signal is maximized while minimizing the power of interference signals (therefore, diagonalizing the combined matrix resulting from the cascade of pre-coding, channel, and combining matrices). In [13], the HBF is achieved by jointly optimizing the DBF and ABF parts in order to comply with a set of constraints

Table 10.1 OFDM numerology for 5G

Band [GHz]	Δf [kHz]	FFT size	Max. active subcarriers	Max. BW [MHz]
0.45–6	15	4096	3300	50
	30			100
	60			200
24–52	60	4096	3300	200
	120			400

(defined by a spectral mask), also minimizing the amplitude variation of the digital signals such that the PA in each RF chain can operate efficiently with a lower input back-off (IBO). However, recent studies indicate that DBF is in fact the most globally energy efficient approach in terms of achievable system throughput per energy consumption unit [11].

The foreseen numerology for OFDM signals in 5G, both in the lower frequency range ($f < 6$ GHz) and the millimeter wave frequency range (24 GHz$< f <$ 52 GHz), are shown in Table 10.1 according to the specifications in [5, 6, 14]. The information included is intercarrier spacing Δf, FFT size, maximum number of active carriers, and maximum occupied bandwidth in both frequency bands. Number of active carriers and bandwidth are expressed in terms of their maximum value because their configuration in 5G NR is flexible and determined by the new parameter bandwidth part (BWP) assigned to the UE [5, 14].

Channel models for millimeter waves take into account the differences in propagation for smaller-wavelength signals, such as: increased scattering and larger penetration losses. These models share some common features with lower frequency channel models (like multipath delay spread). However, they present also fewer and clustered paths, with information on the angles of departure and arrival, as well as increased differences between line-of-sight (LOS) and non-line-of-sight (NLOS) propagation conditions [9, 11].

A compact representation of the channel can be obtained in the beamspace. By means of beamspace representation, it is possible to define an $N_T \times S$ transmit array response matrix $\mathbf{A}_T$ and an $N_R \times S$ receiver matrix $\mathbf{A}_R$ whose columns are steering vectors for a set of S uniformly distributed angles $\theta = [\theta_1, \theta_2, \ldots, \theta_S]$ where $\theta_i = 2\pi i/S$, for $i = 0, 1, \ldots, S - 1$. Then, the $N_R \times N_T$ channel matrix $\mathbf{H}$ can be expressed as a function of the sparse $S \times S$ matrix $\mathbf{H}_S$ as

$$\mathbf{H} = \mathbf{A}_R \mathbf{H}_S \mathbf{A}_T^H, \tag{10.1}$$

where $\mathbf{H}_S$ has p nonzero elements in the position of the angles of arrival and departure, and p is the effective channel length determined by the larger eigenvalues of $\mathbf{H}$.

Considering $\mathbf{A}_T$ as the beam-steering matrix, the angle in which the beam will be directed is determined by the nonzero elements in the corresponding rows of $\mathbf{H}_S$ (and zeros elsewhere), according to the information available on the channel regarding the angles of departure for transmission. Similarly, the information on the angles of arrival in $\mathbf{A}_R$ for combining at the receiver are represented by nonzero elements in the corresponding columns of $\mathbf{H}_S$. The issue is then to design a precoding matrix $\mathbf{U}$ for transmission to obtain the transmitted signal $\mathbf{y} = \mathbf{H}\mathbf{U}\mathbf{x}$, and a combining matrix for reception $\mathbf{V}$ such that the received signal $\mathbf{r} = \mathbf{V}\mathbf{H}\mathbf{U}\mathbf{x} + \mathbf{w}$ is as close as possible to $\mathbf{x}$.

10.1.3 Massive MIMO Challenges

Many issues regarding implementation and open research interests are mentioned for example in [8–10]. Some of these challenges, that motivate research on massive MIMO beamforming systems for 5G, are listed and briefly described below:

- **Efficient signal processing**: Massive MIMO systems will have to efficiently process huge amounts of baseband data. This means that the algorithms used should be as simple as possible to achieve real-time implementation. Linear processing techniques could be a great asset in achieving this goal, as well as distributed coherent signal processing.
- **Hardware impairments**: Transmission over large antenna arrays also means that a great number of RF chains are needed. Thus, the number of up/down converters, PAs, ADCs, and DACs is dramatically increased. In order to keep costs and power consumption at reasonable levels, it is expected that low-cost components should be used. However, these components also present higher impairments that reduce their performance. In particular, high levels of I/Q imbalance and phase noise may be a limiting factor whose impact should be efficiently reduced. In the case of PAs, relaxed linearity constraints to reduce consumption are foreseen and, thus, efficient PAPR reduction algorithms amenable to large antenna systems will be a challenging requirement. Low cost ADCs and DACs present either low-resolution or increased nonlinear distortion, which motivates the design of both compensation algorithms with reduced complexity and processing strategies allowing good performance with low resolution.
- **Channel characterization and estimation**: Many massive MIMO proposals and new results can be obtained based on the assumption that very good channel information is available. Therefore, channel models that accurately reflect the behavior of the channel are key to realistic assessment of system performance. This also motivates research on channel estimation techniques that provides accurate results with reduced complexity. Some trends on channel estimation techniques are: sparse channel estimation using compressive sensing approaches, channel estimation with low-resolution ADCs, and multiuser channel estimation.
- **Reciprocity calibration**: Time division duplex (TDD) will require reciprocity calibration between transmitter and receiver in order to use the same channel estimates for uplink and downlink. However, the cost of implementing it should be carefully addressed in terms of frequency and time resources, additional

components and signal processing. This opens the opportunity to develop new techniques to reduce complexity and increase spectral efficiency.

- **Beamforming for green communications**: Following the pursue of energy efficiency, energy and spectral efficiency co-design is a very interesting topic, particularly in the context of HBF, where an optimum green point of operation can be obtained under certain constraints (varying, for example, the number of antennas and RF chains). This would allow to further boost the power savings offered by HBF architectures.

10.2 Summary of the Book

The presentation of book's contents was divided into three parts, despite that novel contributions were discussed in Part II (Digital compensation techniques in Chaps. 4–6) and Part III (RF imperfection modeling in new technologies in Chaps. 7–9).

Related to *Digital compensation techniques*, power amplifier distortion reduction techniques were discussed in Chap. 4. To this purpose, in addition to introduce figures of merit for in-band and out-of-band distortion, there were discussed transmitter and receiver side compensation schemes. Also, for illustration purposes, a case of study with class AB and ET PAs was included. Chapter 5 is devoted to the analysis of different ADCs. After introducing a characterization of possible distortions, novel (post)-compensation techniques are presented and analyzed by cases of study. Ending this part, Chap. 6 considers carrier frequency offset and phase noise effects in broadband wireless communications. In addition to the analysis of their effects in different applications, downlink and (the more complex) uplink estimation and compensation techniques are studied in detail.

Part III, *RF imperfection in novel technologies*, was started with Chap. 7. There full-duplex communications was studied considering, among other aspects, the system model, the strategies for reduction of self-interference, and the influence of RF impairments in that problem. Chapter 8 follows discussing several aspects of massive MIMO system design. In addition to the basic models for single cell massive MIMO, a study of pre-coding techniques with imperfect channel state information is presented. Furthermore, the analysis of RF minimum requirements in terms of RF and ADC impairments is treated extensively. Finally, Chap. 9 introduces the special design characteristics required for internet of things networks. After a short discussion of proprietary and licensed standards, more detail is put in 3GPP NB-IoT technology. The chapter ends with a study of the interferences that LTE RF impairments could produce in NB-IoT.

References

1. A. Shojaeifard, K. Wong, M. Di Renzo, G. Zheng, K.A. Hamdi, J. Tang, Massive MIMO-enabled full-duplex cellular networks. IEEE Trans. Commun. **65**(11), 4734–4750 (2017)
2. H. Tabassum, A.H. Sakr, E. Hossain, Analysis of massive MIMO-enabled downlink wireless backhauling for full-duplex small cells. IEEE Trans. Commun. **64**(6), 2354–2369 (2016)
3. J. Feng, S. Ma, G. Yang, H.V. Poor, Impact of antenna correlation on full-duplex two-way massive MIMO relaying systems. IEEE Trans. Wireless Commun. **17**(6), 3572–3587 (2018)
4. C. Kong, C. Zhong, S. Jin, S. Yang, H. Lin, Z. Zhang, Full-duplex massive MIMO relaying systems with low-resolution ADCs. IEEE Trans. Wireless Commun. **16**(8), 5033–5047 (2017)
5. ETSI TS 138 211 V15.2.0 Technical Specification, 5G NR Physical channels and modulation, 3GPP TS 38.211 version 15.2.0 Release 15 (2018)
6. S. Parkvall, E. Dahlman, A. Furuskar, M. Frenne, NR: the new 5G radio access technology. IEEE Commun. Standards Mag. **1**(4), 24–30 (2017)
7. M.M. Molu, P. Xiao, M. Khalily, K. Cumanan, L. Zhang, R. Tafazolli, Low-complexity and robust hybrid beamforming design for multi-antenna communication systems. IEEE Trans. Wireless Commun. **17**(3), 1445–1459 (2018)
8. E.G. Larsson, O. Edfors, F. Tufvesson, T.L. Marzetta, Massive MIMO for next generation wireless systems. IEEE Commun. Mag. **52**(2), 186–195 (2014)
9. R.W. Heath, N. González-Prelcic, S. Rangan, W. Roh, A.M. Sayeed, An overview of signal processing techniques for millimeter wave MIMO systems. IEEE J. Sel. Topics Signal Process. **10**(3), 436–453 (2016)
10. S. Kutty, D. Sen, Beamforming for millimeter wave communications: an inclusive survey. IEEE Commun. Surv. Tuts. **18**(2), 949–973 (2016)
11. S. Buzzi, C. D'Andrea, Energy efficiency and asymptotic performance evaluation of beamforming structures in doubly massive MIMO mmWave systems. IEEE Trans. Green Commun. Netw. **2**(2), 385–396 (2018)
12. S. Han, C. I, Z. Xu, C. Rowell, Large-scale antenna systems with hybrid analog and digital beamforming for millimeter wave 5G. IEEE Commun. Mag. **53**(1), 186–194 (2015)
13. V. Venkateswaran, F. Pivit, L. Guan, Hybrid RF and digital beamformer for cellular networks: algorithms, microwave architectures, and measurements. IEEE Trans. Microw. Theory Tech. **64**(7), 2226–2243 (2016)
14. J. Jeon, NR wide bandwidth operations. IEEE Commun. Mag. **56**(3), 42–46 (2018)

Index

© Springer Nature Switzerland AG 2020
F. Gregorio et al., *Signal Processing Techniques for Power Efficient Wireless
Communication Systems*, Signals and Communication Technology,
https://doi.org/10.1007/978-3-030-32437-7

Printed in the United States
By Bookmasters